An alphabetical listing of symbols used in schematics

Note that these are the symbols commonly used in North America and Japan; some other countries use slight variations on these symbols.

Table A-1	Schematic Symbols		
Name/(Abbreviation)	**Symbol**	**Name/(Abbreviation)**	**Symbol**
Antenna		Gate, NAND	
Battery		Gate, OR	
Capacitor (C)		Gate, NOR	
Capacitor (C), Variable		Ground	
Coil		Incandescent lamp	
Coil, Variable		Interconnection, Connected	
Crystal and resonator (X)		Interconnection, Connected 2	
Diode (D)		Interconnection, Unconnected	
Diode (LED), LED		Interconnection, Unconnected 2	
Diode (D), Photo		Inductor (L)	
Gate, AND		Inverter	

(continued)

Electronics Projects For Dummies®

Cheat Sheet

Name/(Abbreviation)	Symbol	Name/(Abbreviation)	Symbol
IR detector		Speaker	
Meter		Switch (S), SPST	
Microphone (MIC)		Switch (S), SPDT	
Motor (M)		Switch (S), DPDT	
Operational amplifier (U or IC)		Switch (S), normally open	
Photocell/Photoresistor*		Switch (S), normally closed	
Piezoelectric buzzer		Transistor (Q), NPN Bipolar	
Power (+V)		Transistor (Q), Phototransistor	
Relay (RLY)		Transistor (Q), PNP Bipolar	
Resistor (R)		Transistor (Q), N Channel Mosfet	
Resistor (R), Variable**		Transistor (Q), P Channel Mosfet	
Solar cell		Voltage regulator (VR)	

*You can use the terms photocell and photoresistor interchangeably.

**A variable resistor is also called a potentiometer.

Copyright © 2006 Wiley Publishing, Inc. All rights reserved.

Item 0968-3.

For more information about Wiley Publishing, call 1-800-762-2974.

For Dummies: Bestselling Book Series for Beginners

Electronics Projects

FOR

DUMMIES®

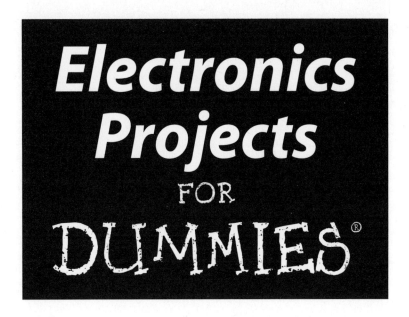

Electronics Projects FOR DUMMIES®

by Earl Boysen and Nancy Muir

WILEY

Wiley Publishing, Inc.

Electronics Projects For Dummies®

Published by
Wiley Publishing, Inc.
111 River Street
Hoboken, NJ 07030-5774

www.wiley.com

Copyright © 2006 by Wiley Publishing, Inc., Indianapolis, Indiana

Published by Wiley Publishing, Inc., Indianapolis, Indiana

Published simultaneously in Canada

For general information on our other products and services, please contact our Customer Care Department within the U.S. at 800-762-2974, outside the U.S. at 317-572-3993, or fax 317-572-4002.

For technical support, please visit www.wiley.com/techsupport.

Wiley also publishes its books in a variety of electronic formats. Some content that appears in print may not be available in electronic books.

Library of Congress Control Number: 2006926111

ISBN-13: 978-0-470-00968-0

ISBN-10: 0-470-00968-3

Manufactured in the United States of America

10 9 8 7 6 5 4 3 2

1B/RT/QX/QW/IN

WILEY

About the Authors

Earl Boysen is an engineer who after 20 years in the computer chip industry, decided to slow down and move to a quiet town in Washington state. Earl is the co-author of *Electronics For Dummies* and *Nanotechnology For Dummies*. He lives with his wife, Nancy, in a house he built himself and finds himself as busy as ever with teaching, writing, house building, and acting. Visit Earl at his Web site to get reviews and information about the latest components and techniques for building projects: www.buildinggadgets.com.

Nancy Muir is the author of over 50 books on topics ranging from desktop computer applications to distance learning and electronics. She has a certificate in distance learning design and has taught technical writing at the university level. Prior to her freelance writing career, she held management positions in the publishing and software industries. She lives with her husband Earl and their benevolent owners — their dog and cat. Nancy's company, The Publishing Studio, has its Web site at www.pubstudio.com.

Dedication

Nancy and Earl dedicate this book to their uncle, Ted Stier, with thanks for being such a great guy and giving Nancy away with such style and grace!

Authors' Acknowledgments

The authors wish to thank Katie Feltman for continuing to hire them to work on interesting book projects and to Chris Morris for managing the editing process and the authors so successfully. Thanks also to technical editor Kirk Kleinschmidt and copy editor Teresa Artman for making sure that what we wrote ended up being accurate and grammatically correct.

We also received help during this project from the following people, and they have our sincere gratitude: Bruce Reynolds of Reynolds Electronics (www. renton.com); the helpful folks at Magnevation (www.magnevation.com); and the following helpful members of our local ham radio club: Clint Hurd, Andy Andersen, Jack West and Owen Mulkey; and Gordon McComb of Budget Robotics (www.budgetrobotics.com).

Publisher's Acknowledgments

We're proud of this book; please send us your comments through our online registration form located at www.dummies.com/register/.

Some of the people who helped bring this book to market include the following:

Acquisitions, Editorial, and Media Development

Project Editor: Christopher Morris

Acquisitions Editor: Katie Feltman

Senior Copy Editor: Teresa Artman

Technical Editor: Kirk Kleinschmidt

Editorial Manager: Kevin Kirschner

Media Development Specialists: Angela Denny, Kate Jenkins, Steven Kudirka, Kit Malone

Media Development Manager: Laura VanWinkle

Editorial Assistant: Amanda Foxworth

Sr. Editorial Assistant: Cherie Case

Cartoons: Rich Tennant (www.the5thwave.com)

Composition Services

Project Coordinator: Patrick Redmond

Layout and Graphics: Claudia Bell, Carl Byers, Joyce Haughey, Barbara Moore, Barry Offringa, Alicia South

Proofreaders: Leeann Harney, Joe Niesen, Christy Pingleton

Indexer: Techbooks

Special Help: **Virginia Sanders**

Publishing and Editorial for Technology Dummies

Richard Swadley, Vice President and Executive Group Publisher

Andy Cummings, Vice President and Publisher

Mary Bednarek, Executive Acquisitions Director

Mary C. Corder, Editorial Director

Publishing for Consumer Dummies

Diane Graves Steele, Vice President and Publisher

Joyce Pepple, Acquisitions Director

Composition Services

Gerry Fahey, Vice President of Production Services

Debbie Stailey, Director of Composition Services

Contents at a Glance

Table of Contents

Introduction

If you've caught the electronics bug, you're ready to try all kinds of projects that will help you develop your skills while creating weird and wonderful gadgets. That's what this book is about: providing projects that are fun and interesting as well as helping you find out about all kinds of electronic circuits and components.

Electronics Projects For Dummies is a great way to break into electronics or expand your electronics horizons. Here, we provide projects that allow you to dabble in using sound chips, motion detectors, light effects, and more. And all the projects are low voltage, so if you follow our safety advice, no electronics folks will be hurt in the process.

Why Buy This Book?

Electronics projects not only help you build useful and fun gadgets, but you pick up a lot of knowledge along the way about how various electronic parts work, how to read a circuit diagram, and how to use tools such as soldering irons and multimeters. So by using this book, you have fun and get some knowledge at the same time.

This book provides you with just what you need to get going in the fun world of electronics. It offers projects that you can build in a reasonable amount of time — and in most cases, for under $100 each (some well under!).

Foolish Assumptions

This book assumes that you have an interest in electronics and that you've probably explored the world of electricity and electronics a bit. You've probably scanned a few electronics circuit Web sites and maybe a magazine or two and have picked up some of the jargon. Other than that, you don't need anything but a minimal budget to buy parts and tools, a small space in your house or apartment that you can set aside for a workbench, and a little time.

If you feel like you want more information about terms and concepts in electronics to help you out, we recommend *Electronics For Dummies,* by Gordon McComb and Earl Boysen (Wiley).

You don't need to be an electrical engineer or have worked on electronic projects in the past. We provide some initial chapters that help you stock up on essential parts and tools, understand what each one does, set yourself up for safety, and master a few simple skills. Then you're all set to tackle any one of the projects in this book.

Safety, Safety, Safety!

We can't say this enough: Electronics, especially lower-voltage projects like the ones in this book, can be a painless pastime but only if you follow some basic safety procedures from the get-go.

Even low voltages can harm you, soldering irons can burn you, and small pieces of plastic or wire that you snip could fly into your face.

We recommend that everybody — even those with electronics experience — read the chapter on safety (Chapter 2). And because we can't cover every potential danger in a single chapter, be sure to read each manufacturer's warnings about how to use parts, power sources, and tools. Finally, use common sense when working on projects. If in doubt whether a safety precaution is necessary, just do it. *Better safe than sorry* is one of our mantras.

How This Book Is Organized

Electronics Projects For Dummies is organized into several parts, starting off with some general information about safety and stocking your electronics workshop. Then we offer several parts with different types of projects, and finally conclude with the Part of Tens chapters with additional resources you might want to explore. This book also has a spiffy full-color photo spread of some of the circuits and finished products of several of the projects.

Here's the rundown of how this book is organized.

Part I: Project Prep

If you're new to electronics, read through this part first. Even if you're seasoned, humor us and read Chapter 2 about safety. Then use Chapters 3 and 4 to gather the parts and tools you'll need and also bone up on some essential electronics skills, such as soldering and reading schematics.

Part II: Sounding Off!

This part contains the first set of projects, all involving sound in some fashion. Here you work on projects to make lights dance to music, create a parabolic microphone to pick up sounds at a distance, make a wizard that talks when you push his buttons, and create your own AM radio.

Part III: Let There Be Light

Electricity can produce light (as Thomas Edison could have told you), so here we show you how to work with light in a variety of ways. These projects use light to amuse or even make gadgets run. In this part, you light up a pumpkin by using a motion detector, create a light display that will make your next party rock, and build a go-kart that you direct by using an infrared remote control device.

Part IV: Good Vibrations

Some electronic gadgets do their thing when they sense vibrations. All the projects in this part depend on vibrations, including electrical, mechanical, or radio waves. Work through these projects to create a metal detector, a radio controlled vehicle that senses light and runs around a track, and a device that sits on your couch and raises a ruckus if your pet jumps on the cushion.

Part V: The Part of Tens

The chapters in this part provide the ever-popular *For Dummies* top-ten lists. Use the recommendations here to explore some interesting suppliers of electronic parts and tools; get information or swap ideas about general electronics topics online or in print; or look into resources for more specialized interests, such as audio effects and robotics.

Icons Used in This Book

We live in a visual world, so this book uses little icons to point out useful information of various types.

The Tip icon points you to information that is interesting and can save you time or headaches. These icons generally add a bit of spice to your electronic project education.

Oops. If you don't heed these little icons, you might regret it. Warnings alert you to potential danger or problems that you want to avoid.

Remember icons remind you of an important idea or fact that you should keep in mind as you explore electronics. They might even point you to another chapter for more in-depth information about a topic.

If you're gonna build an electronics project, you're gonna spend some money. To save you time and help you keep your costs down, we give you shopping tips wherever you see this icon.

Part I
Project Prep

The 5th Wave   By Rich Tennant

"So I guess you forgot to tell me to strip out the
components before drilling for blowholes."

In this part . . .

Before you can jump in and tackle projects, you might want to brush up on (or discover for the first time) the basics. Chapter 1 answers such urgent questions as "What is an electronics project, anyway?," and Chapter 2 provides our best advice about safety procedures that keep you intact while you play with gadgets. Chapter 3 runs down the parts and equipment you work with in a typical project, and Chapter 4 reviews some basic skills that you need to build all kinds of electronic toys.

Chapter 1

Exploring the World of Electronics Projects

In This Chapter

▸ Understanding exactly what an electronics project is

▸ Exploring the effects you can achieve

▸ Considering what's in it for you

▸ Determining what you need to invest to get started

Y ou probably picked up this book because you love tinkering with gadgets, from that train set you got as a kid to the motion-activated dancing monsters on display in the store aisles at Halloween. Not only are you intrigued by them, but you wonder whether you can build something like them yourself. Now that you own this book, yes, you can!

In this chapter, we take a look at exactly what getting into building electronics projects involves, the kinds of great gadgets you can build yourself, what you'll get from spending your time with electronics, and what you need to commit to take the plunge.

What Is an Electronics Project, Anyway?

Obviously, an *electronics* project involves electronics, meaning that you use electricity to make something happen. However, overlaps exist among electronics, mechanics, and even programmable devices such as robots. Here's what we mean when we say *electronics projects*.

Electronics, mechanics, robotics: Huh?

Do you dream of building elaborate Erector Set-types of mechanical structures — perhaps a model of the Golden Gate Bridge with pulleys and levers moving objects around? Is your goal to create a robot butler with a programmed brain that enables it to serve your every whim? Well, those aren't exactly what we categorize as electronics projects.

Certainly, electronics projects are often combined with mechanical structures that use motors, and a robot has electronic components driven by microcontrollers and computer programs. In this book, though, we focus on projects that use simple electronics components to form a circuit that directs voltage to produce effects such as motion, sound, or light. By keeping to this simple approach, you can pick up all the basic skills and discover all the common components and tools that you need to work on a wide variety of projects for years to come. For these projects, you don't have to become a mechanical or programming whiz.

An electronic circuit might run a motor, light an LED display, or set off sounds through a speaker. It uses various components to regulate the voltage, such as capacitors and resistors. A circuit can also use integrated circuits (ICs), which are teeny, tiny circuits that provide a portion of your circuit in a very compact way. This saves you time micromanaging pieces of the project because somebody else has already done that job for you, such as building a timer chip that sets off a light intermittently.

Programmable versus nonprogrammable

ICs are preprogrammed or programmable. And that brings us to our next distinction.

Although we do use ICs in many of our projects — for example, in the form of a sound chip that's preprogrammed with beeps and music — for the most part, we keep away from programmable electronics. In order to work with programmable electronics, you have to get your hands dirty with programming code and microcontrollers, and that's not what we're about here. Instead, we focus on building electronics gadgets that teach you about how electricity works and get your mind stirring with ideas about what you can do by using electronics, rather than computers.

Don't get us wrong: Microcontroller projects can be a lot of fun. After you get your hands dirty and pick up lots of basic skills doing the projects in this book, you might just go out and buy *Microcontroller Projects For Dummies* (if such a book existed).

Battery-powered versus 120 volts+

One other thing that we made a conscious decision about when writing this book was that we didn't want you tinkering with high-voltage projects. Electricity can be dangerous! Keeping to about 6 volts keeps you reasonably safe whereas working with something that uses 120 volts — like the juice that comes out of your wall socket — can kill you. While you're discovering the basics of electronics, our advice is that it's better to be safe than sorry.

When you get more comfortable and more knowledgeable about tools and skills and safety measures (which we put a lot of emphasis on, especially in Chapter 2), you might explore higher-voltage projects such as high-powered audio or ham radio projects. In this book, we show you how to work with low-voltage batteries and still have fun in the process.

Mixing and Matching Effects

The possibilities of what electronics projects can do are probably endless; on a basic level, the projects in this book use electricity to do a variety of things, from running a small cart around the room to setting off a sequence of lights or sounds.

Generally, most electronics projects consist of four types of elements:

- ✔ **Input:** This sets off the effect, such as a remote control device or a switch that you push. An event and a sensor, such as a motion or light detector, can also be used to activate an effect.
- ✔ **Power source:** We typically use batteries in these projects.
- ✔ **Circuit:** Components that control the voltage — such as transistors, capacitors, amplifiers, and resistors — are connected to each other and to the power source by wires and make up the circuit.
- ✔ **Output:** This is what is powered by the circuit to produce an effect, such as speaker emitting sound, LED lights going off, or a motor that sets attached wheels spinning.

What Can You Do with Electronics Projects?

You get to explore a number of variations in the projects in this book. And sure, this stuff sounds like it might be cool, but what's in it for you? Electronics projects offer three benefits (at least):

- ✔ Fun
- ✔ The thrill of making something work all by yourself
- ✔ A boatload of useful knowledge

Just for the fun of it

One obvious benefit of tinkering with gadgets is that it's just plain fun. If you're the type who's intrigued by how things work and what's under the hood, you probably already know this.

In fact, we have lost ourselves for hours figuring out circuits (this is the electronics equivalent of a jigsaw puzzle, which starts as a drawing, like the one shown in Figure 1-1), wiring the components, and refining the results. You can also, quite literally, amaze your friends with the things you build. And if you go in for electronic gizmos that you can race, scare people with, or use to entertain crowds at parties, you can share the fun with others.

Don't forget the social aspect: Electronics projects devotees comprise a friendly bunch of folks who like to help each other. You can get into discussion groups online or join a local electronics club and find both interesting ideas and friendships at the same time. Chapter 16 provides ten great Web sites about electronics where you'll find such online groups.

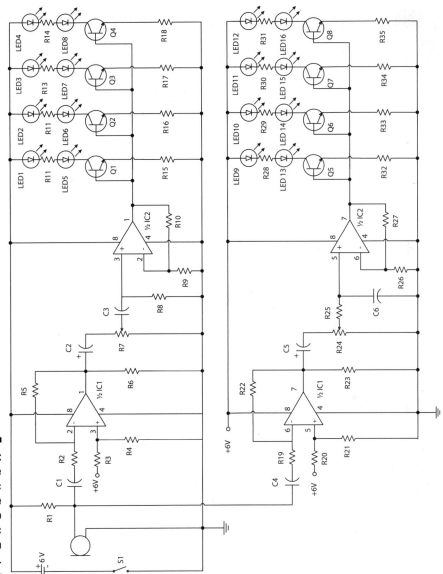

Figure 1-1:
The
schematic
for the
Dance to
the Music
project in
Chapter 5.

Building things you can actually use

So why, when you can buy an AM radio for $7.95, would you decide to build one yourself with parts that cost $30? That's a good question. The truth is just about everything you build in the projects included in this book — and most of the circuits floating around on the Internet — is something that you could probably buy in some form somewhere. But where would the challenge be in that?

Here's why hundreds of thousands of electronics junkies build instead of buying: Because they can. They can make something that grabs music out of the airwaves or sets off a light display or sends a little cart wheeling around the room themselves. We guess this is why people knit sweaters instead of buying them or work on old cars instead of taking them to mechanics. It just feels good to master something on your own.

Parts II, III, and IV of this book are where you can find all these cool projects, divided into categories by what the projects do, such as producing light, sound, or motion.

Some of the things that you build in this book are just for fun, like the dancing dolphin light display (Chapter 10). Other things have a practical use: the Couch Pet-ato (Chapter 14) keeps your cat off the furniture when you leave the house, for example.

Besides building gadgets that have a use, in some cases, you can build items more cheaply than you can buy them in the store. You could just end up with projects you can put to work and save a few bucks in the process.

Picking up lots of cool stuff along the way

One of the great things about electronics is that it teaches you about all kinds of things you can use in your life. For example, you discover

- ✔ How electricity works and how to stay safe when working with it
- ✔ How to read an electronic circuit and build it on a breadboard like the one shown in Figure 1-2
- ✔ How to use a variety of tools to solder, build, and customize casings to hold your gadgets
- ✔ How to work with integrated circuits
- ✔ A bit about wiring (which can give you a head start when you decide to learn how to add an outlet to your kitchen someday)

Figure 1-2:
Here's
what the
breadboard
for Dance to
the Music in
Chapter 5
looks like.

This book is full of lots of School of Hard Knocks information that might take you years to acquire doing electronics projects on your own; you'll also pick up lots of wisdom as you work through the projects and try things out for yourself.

What You Need to Get Started

Now that you're all excited about the benefits of working on electronics projects, you're probably wondering what this will cost you in dollars and workspace.

How much will it cost?

We tried to keep the cost of the projects in this book to under $100; in many cases, the materials and parts will cost you under $50 or so.

Depending on what you have lying around the house already, you might not have to invest in some of the basic tools, such as pliers or a screwdriver. You will probably have to spend $50 or so for electronics-specific tools and materials such as a soldering iron, solder, and a multimeter like the one shown in Figure 1-3.

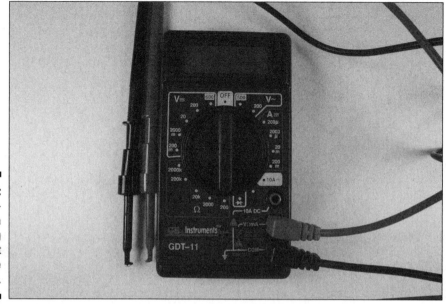

Figure 1-3:
A multi-meter is a measuring device that you'll use often.

If you want to get really fancy, you could spend a couple hundred dollars on fancy testing equipment such an oscilloscope, but you don't have to have that equipment to get through these projects, by any means.

Of course, in the world outside this book, projects can cost you hundreds of dollars. Like any hobby, you can spend a few bucks to dabble or mortgage your house to get into it in a big way. To get your feet wet in electronics, though, the investment is not that great.

Keep in mind that you can reuse some of the parts of one project (such as a breadboard) in another and cut your electronics budget further.

See Chapter 3 for information about the parts and tools that we recommend you get to build your basic electronics workshop.

Space . . . the final frontier

One thing you do need to leap into the world of electronics projects is space. That doesn't mean you have to take over your living room and build a fancy workbench. In most cases, a corner of your garage or laundry room stocked with a shelf where you can keep parts and a card table works just fine. We do advise that you find a specific space for your projects.

In short order, your workspace will be filled with tools and parts and all kinds of (useful) junk (see Figure 1-4). See Chapter 2 for advice about safety when working with all this stuff. For example, stock your workspace with safety glasses that protect you whenever bits of wire go flying, and find a place where you can keep your soldering iron in a stand so it doesn't roll into your lap.

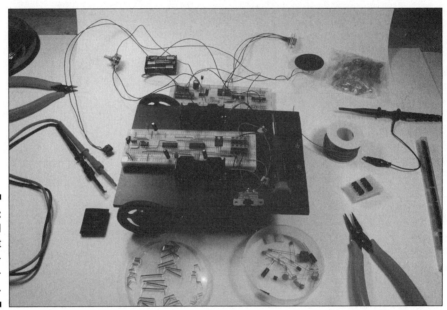

Figure 1-4:
A typical assortment of electronics para-phernalia.

We also recommend finding a spot that you can close off if there are others in your household — especially small children or pets — who could topple your work surface or eat tiny electrical parts and do themselves damage. Electronic projects don't happen in a day, and you might work on a single project over a matter of weeks. If you have a small room with a door to keep others out, great. If not, use your common sense about what you leave out on your work surface overnight.

Chapter 2

Safety First

. . . .

In This Chapter

▶ Avoiding those nasty shocks

▶ Keeping your electric components safe from static discharge

▶ Working safely with tools

▶ Keeping yourself and your workspace neat and tidy (and safe)

. . . .

*W*e won't kid you: ***Electricity deserves your respect.*** It can shock you, burn you, and even kill. In this book, we stick with projects that work with AA batteries to limit the potential for serious damage.

Still, anytime you work with electronics, there is potential for danger. If these projects get you excited about electronics so that you move on to projects that use bigger jolts of electricity, now is the time to learn the proper respect for electricity and the proper safety precautions when working with electronics projects.

In this chapter, you discover what electricity is capable of — and how to keep yourself, electrical components, tools, and those near and dear to you safe.

This is the one must-read chapter in this book. Humor us, and read it from top to bottom, okay?

Avoiding Shocks Like the Plague

Your body is a delicate machine. Electric shocks, depending on certain conditions, can be fatal, even at relatively low voltages. What comes out of your wall outlet is deadly if you play around with it. Even electrical gadgets working off batteries can cause you serious damage.

How voltage and current can get you

Your body is like a big resistor. Usually, your body's resistance is high enough to prevent damage when you're exposed to low voltages. However, certain conditions can lower your body's resistance, lowering the amount of voltage needed to cause you serious damage, such as giving you a nasty burn. Those conditions might include handling electronics with sweaty palms or trying to change your 12 volt (V) car battery on a rainy day — either can turn a slight tingle into a fatal event.

Both *AC* (alternating current, such as the power from your wall outlet) and *DC* (direct current, such as from a battery) voltage can damage you in different ways:

- ✔ **AC voltage:** This type of voltage regularly reverses direction. This can cause your heart to shift its regular beating pattern in a condition known as *ventricular fibrillation*. If this happens, your heart muscles go out of whack in a way that causes blood to stop pumping. In this situation, even if you cut the current, your heart might not be able to find its proper rhythm, and you could die.

- ✔ **DC voltage:** This type of voltage is on constantly and causes your muscles to contract and seize up quickly (including your heart muscle). If you grab an electrical device in conditions that cause your body to conduct DC voltage, your hands could become frozen (unable to let go of the device), and your heart could stop. If someone cuts the current quickly, though, your heart might begin to beat again (and you'll be able to attend that Rotary luncheon next week).

Short of killing you, electric shock can cause burns as the current dissipates across your body's natural resistance (that is, your skin).

How much is too much?

Most resistance in your body is in your skin. If your skin is wet or damp, that resistance is lowered. If you handle an electrical device with damp hands, even voltages under 20V or so (not enough to even light a low-wattage lamp) might be sufficient to do you serious damage. The 120V coming out of your electrical outlet has a lot of punch: more than enough to kill you.

Four AA batteries in series — which is what we use in the projects in this book — generate only about 6V. We did that on purpose to keep you relatively safe.

Is it the voltage or the current — or both?

Electricity is the movement of electrons *(current)* through a conductor when a voltage is applied across the conductor. Electric current is what burns your skin, stops your muscles cold, and causes your heart to go into fibrillation. If you touch a *live* wire (that is, any conductor at some voltage), current can flow through your body because it is a conductor. The amount of current that flows through your body depends upon your body's resistance to the flow of electric current and how much voltage is applied.

Ohm's Law deals with the relationship between voltage, current, and resistance. Here's the law, for those among you who appreciate equations:

Current = Voltage ÷ Resistance

The calculation for what's dangerous involves the current, the voltage, and your body's resistance. The current passing through your body is equal to the difference in the voltage that's being applied to two spots on your body (for example, your hand touching an electric circuit and your feet touching the floor, or one hand touching a live conductor and another hand resting on a chair), divided by your body's resistance.

Just because AA batteries don't have a high voltage output, don't think that they can't hurt you. If you short them out, all the electrons will flow quickly from the negative to the positive poles and generate a lot of heat — enough heat, in some cases, to destroy the battery and possibly burn you. If you feel heat coming from your circuit or the batteries, you might have a short-circuit or a component inserted the wrong way. Turn it off and let things cool down; then check to see what's causing the problem.

The resistance in your body can vary greatly. For example, if you have sweaty hands and touch a live wire with one hand while the other hand rests on a metal table, this is a very dangerous situation. Because you have moisture on your hands — which lowers your contact resistance — a higher current will flow through your body for a given voltage. If you have dry hands — one hand touching a live wire, the other hand in your pocket — and your feet on a dry, rubber mat, there's far less danger from the same amount of voltage because your resistance is higher. However, if a higher voltage comes your way, even with the higher resistance, you could die. Bottom line: There is no iron clad rule as to what level of voltage will kill or seriously injure a person because of all the variables.

Regardless of how much voltage you work with, develop safe work habits now.

Common sense: Protecting yourself from getting shocked

Although you should always use care working with electricity, we want to let you know some common situations to avoid that could turn your body into a super conductor. You know you shouldn't stick your finger into an electrical outlet (we hope!), but you should also get into some other good habits. Read on.

Rings are out

Metal is a dandy conductor. Wearing rings or other metal jewelry around electricity is a lousy idea. For example, when the skin on your finger is surrounded by a ring (a terrific contact point) and you touch a voltage source, your body's resistance can be very low. In that state, even a lower voltage jolt could do you serious damage. Leave jewelry somewhere else. (Tell your spouse or fiancée that we said it's okay for you to take off your wedding ring when working with electricity.)

Another good reason to avoid jewelry is that it can snag on things. Imagine working on a breadboard filled with wires and tiny components, only to have your ring or necklace catch on something and yank it out. At the least, you have to put the component back in place; at the worst, you could damage the component and have to replace it.

Beware of water!

Don't work in a wet environment (say, outdoors on a rainy day, or while standing on a damp garage floor). This prevention might seem obvious, but consider that cup of coffee on your workbench. What would happen if you knocked it over while working with electricity? You need to become super careful about *anything* wet or moist in or near your work area. This includes you: If you just came in out of the rain or from a run, dry off before working on electrical equipment.

Respect electricity

Here's one simple rule that you should memorize right now: Never touch a component in a circuit that has power (an *energized circuit*). Turn off all power sources or remove the source from the circuit entirely before touching it.

One trick that electricians use is to keep their left hands in their pockets when working with electrical equipment. If a zap occurs, it will flow from their right hands to the ground — not from hand to hand, passing right through the heart. You shouldn't be working with live electricity — ever! — but this trick of trade used by more advanced users shows how important it is to understand how electricity works and respect its authority.

Even when the source is removed, some electricity might remain. To be absolutely sure, before you touch anything, test the circuit with your multimeter. (We talk about how to use a multimeter in Chapter 4.) And don't take somebody else's word that the power is off; always check and double-check this yourself!

 Don't work with AC-operated circuits unless you absolutely have to. And if you do, it might not be a bad idea to have a friend nearby who is trained in CPR. Visit www.redcross.org for more information about CPR training.

Protecting Electronic Components from Dreaded Static Discharge

You're not the only thing in your work area that could suffer from shocks. Static discharge (also referred to as electrostatic discharge; ESD) can do damage to your delicate electrical components. Static discharge is so named because it's caused by the discharge of electrons from a static charge that hang around in an insulating body, even after the source of those electrons goes away.

Static charge is typically caused by friction. You might trap some electrons in your body as you walk across a carpet, for example. When a static charge is built up on your body, a corresponding voltage difference is built up between your body and a grounded object, such as a doorknob. The zap when you then touch a doorknob is the static discharge: that is, the electrons flowing from you to the doorknob.

What static discharge can do

Metal oxide semiconductor (MOS) devices are cool because they allow integrated components to use less power. MOS devices improve circuit design and operation, but that improvement comes at a price. These little guys are VERY sensitive to ESD. One little zap, and they are likely to be history.

When you walk across a carpet, you can produce a static charge in the range of 2,000–4,000V. Because the number of electrons trapped on your body is low, you feel only a little shock. However, MOS devices contain a very thin layer of insulating glass that can become toast when exposed to as little as 50V of discharge or less. When you work with a MOS device, your body, clothes, and tools have to be free of static discharge. (You find out how to do that in the next section.)

MOS devices are found in many integrated circuits (ICs) and transistors. ICs and transistors that use bipolar devices do not have the very thin layers of insulating glass found in MOS devices, so they are less susceptible to damage from static discharge. Resistors, capacitors, diodes, transformers, and coils, on the other hand, aren't in too much danger from static discharge. Keep static discharge away from your projects just to be safe.

How to guard against ESD

To get rid of static discharge in your electronics workshop, you can do several things, such as wearing anti-static devices and clothing, using static-dissipative floor mats, and grounding your tools.

First, wear an anti-static wrist strap. An anti-static wrist strap is one of the best ways to get rid of ESD. This strap, like the one shown in Figure 2-1, fits snugly on your wrist. You then attach the wire on the strap to *earth ground* — which is just what it sounds like: namely, the earth beneath your feet.

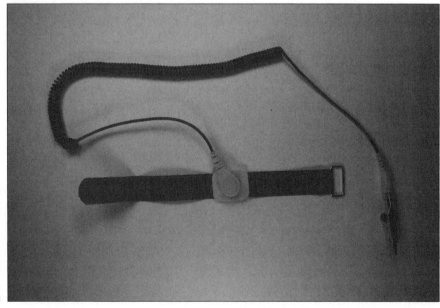

Figure 2-1:
An anti-static wrist strap is ESD's worst enemy.

The cold water pipe on a water heater or under a sink is a good option for earth ground — if the water pipes are metal, that is. Plastic water pipes that you find in some newer construction won't work. Because the cold water pipe comes up out of the ground, it is therefore grounded (logical, huh?), which works where the hot water pipe usually won't. Use a clamp to attach a wire to the pipe (earth ground) and run it to your worktable, being sure to run the wire along the wall so you don't trip over it. Set a loop of the wire at the edge of your worktable where it's handy to attach the alligator clip on the end of it to your wrist strap.

If you don't happen to have a metal cold water pipe nearby, the best method is to use a metal rod that you insert into the ground. The standard rule is to sink it three feet deep.

Second, wear clothing that is less likely to accumulate static charge. For example, polyester, acetate, and wool fabrics easily accumulate static charges whereas as cotton is less likely to accumulate the static charges necessary for ESD.

Using an anti-static wrist strap and wearing cotton clothing will usually be sufficient.

Third, if you plan to do electronics projects long-term, consider buying a static-dissipative mat for your work surface. You connect the mat to a ground, as you do with the wrist strap, and the mat dissipates charges from components you're working on as you lay them on the mat. However, the mat has a high enough resistance that it won't short together the pins of components.

There are also static-dissipative floor mats; however, these are more likely to be used in a manufacturing setting when a worker needs to move between workstations.

Anti-static wrist straps and static-dissipative workbench mats can be purchased at most electronics distributors. See Chapter 15 for a list of electronics distributors. The prices for wrist straps vary widely but start at just over $6; prices for workbench mats start at about $10.

Don't try rigging up a homemade anti-static wrist strap. The ones you buy have a high resistance that slowly dissipates current. If you use a material without that resistance, the current would rush to ground — which could cause you serious injury — instead of slowly dissipating. For $6, why take a chance?

Finally, don't forget to ground your tools. Some tools, like the better soldering irons, have a three-prong plug that provides a ground connection. (Cheap tools might use only two-prong plugs, so avoid them at all cost.) Other than a grounded soldering iron, however, most metal tools (such as a screwdriver) dissipate static through you when you wear your handy anti-static wrist strap.

Working with the Tools of the Trade

In addition to keeping yourself safe from electricity, you will find a few other dangers with working with electronics projects. Using a variety of tools — from a hot soldering iron to a sharp hacksaw — mandates that you adopt some wise safety habits.

Safe soldering

Soldering poses a few different dangers. (You might use solder to attach various pieces of your electronics project, such as soldering wires onto a speaker, microphone, or switch.) The soldering iron itself (you can see one in Figure 2-2) gets mighty hot. The *solder* (the material you heat with the iron) gets hot. Occasionally, you even get an air pocket or impurity in solder that can pop as you heat it, splattering a little solder toward your face or onto your arm. To top that off, hot solder produces some nasty fumes.

Figure 2-2: Always treat a soldering iron with respect!

Soldering itself takes experience to get right. Your best bet is to have somebody who is good at it teach you.

Here are some soldering safety guidelines you should always follow:

- Always wear safety glasses when soldering.
- Never solder a *live* circuit (one that is energized).
- Soldering irons come in models that use different wattages. Use the right size soldering iron for your projects, as we discuss in Chapter 3; too much heat could ruin your board or components.

✔ Solder in a well-ventilated space to prevent the mildly caustic and toxic fumes from building up and causing eye or throat irritation.

✔ Always put your soldering iron back in its stand when not in use. Too, be sure that the stand is weighted enough or attached to your worktable so that it doesn't topple over if you should brush against the cord.

✔ NEVER place a hot soldering iron on your work surface. You could start a fire.

✔ Give any soldered surface a minute or two to cool down before you touch it.

✔ Never, ever try to catch a hot soldering iron if you drop it. No matter how hard you try, you are very likely to grab the hot end in a freefall. Let it fall; buy a new one if you have to — just don't grab!

✔ Never leave flammable items (like paper) near your soldering iron.

✔ Be sure to unplug your soldering iron when you're not around.

Don't put your face too close to the soldering site because of the danger of stray hot solder and those horrible fumes. Instead, use a magnifying device to see when soldering teeny-tiny components to a board. You can buy clamp-on magnifiers that keep your hands free for soldering.

Running with sharp objects: Cutting, sawing, and drilling

As you work with electronics projects, you will find yourself spending a certain amount of time doing construction tasks: building enclosures of various shapes and sizes, cutting holes for switches, drilling a board to attach wheels, and so on. These tasks involve using tools such as knives, saws, and drills.

Anything that cuts can cut you, too. Here are a few tips for safe cutting:

✔ **Take a moment before you cut.** Know where you want to cut, what the best tool for the cut is, and how best to hold onto the thing you're cutting to avoid cutting your fingers. (Clamps or a vise are useful for securing whatever you are cutting.)

✔ **Get experience.** If you're new to sawing and drilling, get an experienced hand to fill you in or take a shop class.

✔ **If you don't know how to run power equipment, don't use it.** A small, unpowered hand tool can often perform the job without as much potential danger to you if something goes wrong.

✔ **Keep distractions to a minimum.** If you're likely to have a visitor wander in while you're running a power saw, put a Do Not Disturb sign on the door. That momentary distraction could cause an accident.

✔ **Don't hurry.** When you're rushed, you make mistakes and accidents happen.

✔ **Never force things.** If the drill is meeting resistance or the saw isn't biting into the material, stop and check out the situation. Forcing a tool at these times can cause it to kick back on you or worse.

✔ **Wear leather work gloves to avoid cutting your hands** when handling materials with sharp edges or a rough surface that could have splinters.

✔ **Watch what you wear.** Always wear safety glasses when cutting with any tool to avoid flying bits landing in your eyes. If a power tool is noisy, protect your ears with ear muffs or ear plugs. (See the last section of this chapter for more about this.)

✔ **Your safety rules apply to anyone in your work area.** A flying object could hit a friend in the eye several feet away, and a noisy tool could damage his hearing.

✔ **Keep a first aid kit handy, just in case.** Taking a first aid class would also be a good thing.

✔ **Have a phone handy for emergencies.**

✔ **Follow directions!** Power tools often have safety devices and usually come with instructions for their use. Always engage the safety devices and follow the manufacturer's recommendations for safe use of the tool.

In the hopes that you'll rush right over to the tool aisle and buy a lot of tools, many home improvement stores offer free classes on using power tools and other procedures that might help you get started.

A Safe Workspace Is a Happy Workspace

The environment that you work in can be as important to electronic project safety as how you deal with electricity or sharp tools. Paying attention to details, such as the kind of clothing you wear — as well as how neat and tidy you keep your work space — pays off by reducing mistakes and accidents.

Dressing for safety

We put things you wear around your workshop into two categories: the clothing you come in with and the safety devices you should put on as you work.

The clothes make the man (or woman) safe

Here are two important considerations for safety in the clothing that you put on before going into your workshop. We touch on each of these elsewhere in the chapter, but they bear repeating:

✔ **Don't wear loose-fitting clothing.** Loose-fitting clothing and items like scarves or ties can get caught on tools or other items. This could cause you to get a burn, have a fall, or knock a sharp object off your workbench. Wear comfortable clothing — just not clothing that flaps around. In fact, humor us and tuck in that shirttail right now, okay?

✔ **Wear the right fabric.** Fabrics made of cotton don't hold static charges as easily as man-made fibers do. Static discharge can zap electronic components into oblivion. Leave the polyester leisure suit in your closet, and opt for the cotton jeans and shirt instead.

Arming yourself for safety

You should put on certain safety devices — such as ear protection, safety glasses, and leather work gloves — depending on the kind of work you're doing.

Ear protection makes sense if you're working with loud noises, such as when running a very loud power tool. With small electronics projects, like the ones in this book, you probably won't use a very loud piece of equipment. But if you graduate to working on life-size robots, consider your hearing when working with power tools. You can purchase ear muffs, like the ones shown in Figure 2-3, to protect yourself.

Figure 2-3:
Safety ear muffs to protect your hearing.

As Mom used to say, when they were handing out eyes, you get only two, so take care of them. Safety glasses, like those shown in Figure 2-4, are practically a religion with us. In fact, we'd almost go so far as to say when you enter your workspace, put on safety glasses. That could be going too far, but there are many, many instances when you should wear them: when cutting anything, soldering, clipping wires, and so on. Consider whether safety glasses wouldn't be a good idea before you do any procedure.

Don't delude yourself that regular prescription glasses will protect you. They aren't necessarily made of shatterproof material. Too, they have no protection along the sides.

Figure 2-4:
Our very favorite electronics project prop.

Direct things that you're cutting down — toward your workbench — instead of up toward your face. As long as you can make the cut without looking, this can guarantee that flying pieces go away from you and not toward you.

If you're working a lot with something that generates fumes (anything from paint to solder), consider taking a cue from Zorro and wearing a mask called a *respirator* (but wear yours over your mouth and nose) that can be found at any hardware store. Respirators are rated for different types of protection, so make sure you get the appropriate one. For example, one type might keep small particles like sawdust out of your mouth, and another might be designed to keep fumes at bay.

Electronics and alcohol don't mix

Okay, this probably goes without saying, but don't work around electronics if you've had a drink (or two or three). Alcohol slows your brain and has an impact on your judgment. Bad judgment around electricity or sharp power tools could be fatal. 'Nuff said?

Clean up your stuff!

Keeping your workspace neat, including minimizing strung-out cords that could trip you up, is important in preventing accidents.

A cluttered work surface makes it hard to see what you're reaching for. You might reach over to grab a plastic box, only to come up with a mini hacksaw in your palm (ouch!).

Pick up small pieces of cut wire or loose screws and nails. Not only could you step on one and cut yourself if you decide to work barefoot someday (not something we recommend), but a pet or child could pick up such small items and decide they would be tasty.

Keeping kids and pets out of your space

Besides keeping your space neat, you should also keep your space off-limits to the smaller members of your household: namely, kids and pets. Even if you put away all sharp tools religiously and make sure that any power source is disconnected from breadboards, small hands (or paws) are made for mischief.

If you can, lock your workspace when you're not in it. If you can't (maybe because your workspace is a corner of your den), lock up your electronic project tools, components, and works in progress in a box or cabinet.

Chapter 3

Assembling Your Electronics Arsenal

In This Chapter

▶ Gathering the right tools for the job

▶ Collecting electronic components to make projects run

▶ Finding what you need to build what surrounds your project

▶ Building circuits on breadboards

*W*hen you meet somebody who has had a hobby for a few years, he or she usually has a well-stocked arsenal of materials and tools for the task at hand. Knitters have drawers full of wool; stamp collectors have tweezers and scrapbooks; and electronics people have drawers full of switches, resistors, capacitors, integrated circuits, and transistors.

In this chapter, we walk you through the typical items that you need to work with for electronics projects. We introduce you to just about all the tools and components and building blocks that you use in the projects in this book. Whether you buy some now or wait until you need them, by the time you finish this chapter, you will be familiar with the most common tools of the electronics trade.

Tool Time

Many of the tools we want to first address are those that you find in the tool aisle of your hardware or home improvement store — everything from the somewhat specialized soldering iron to the ubiquitous screwdriver.

Soldering prerequisites

If you've ever used wax to seal an envelope, you understand the basic premise of soldering. Take a material (in this case, solder; pronounced *sod*-der) and heat it so that it melts onto items, such as two wires you have twisted together to make a physical connection. When the solder cools, you have a seal or joint that makes an electrical connection between the items.

Soldering requires that you get your hands on a few basic items:

- ✔ **Soldering iron:** See an example of one in Figure 3-1.

   Get one rated at about 30 watts, preferably one for which you can buy different size tips so you can work on different types of projects. And make sure to get an iron with a three-prong plug so that it will be grounded.

- ✔ **Tips:** Large tips can be chisel-shaped and about ⅛" wide; small tips can have a cone shape with a radius at the tip of only ¹⁄₆₄". Most soldering irons don't specify the tip sizes that are supplied with the iron. For most electronics work, we suggest you just find one described as a fine tip at a electronics supplier. If you're ordering a replacement tip, a ³⁄₆₄" cone shape is a good size for general use. If you're soldering circuit boards, you might try a ¹⁄₆₄" cone-shaped tip. Figure 3-1 shows a soldering iron with a collection of different tip sizes and shapes.

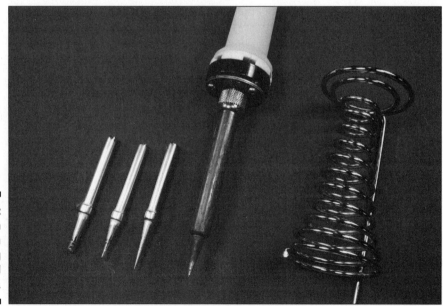

Figure 3-1:
A collection
of tips, a
soldering
iron, and
a stand.

If you end up doing a lot of projects soldering components on circuit boards, you might decide to spend extra — sometimes quite a lot extra — to get a soldering iron with controls that allow you to change the wattage, or even one that senses the temperature of at the tip of the soldering iron and adjusts the power to keep the temperature stable.

✔ **Stand:** You need a device to hold the soldering iron. To ensure that it doesn't tip over with a hot soldering iron in it, make sure that the stand's base is heavy enough or that you can clamp it to your worktable.

✔ **Damp sponge:** You will use this constantly to clean the soldering tip between soldering jobs.

✔ **Solder wick:** This piece of flat, woven copper wire — also called a *solder braid* — soaks up solder when you need to rework a connection and need to remove a dab or two from a joint. Some folks use a desoldering pump to suck up solder, but we find that a wick is easier to use.

✔ **Solder:** *Solder* is a material that when heated and then cooled, holds wires and other metallic connectors together. The standard type used for electronics is referred to as *60/40 rosin core* solder, which is 60 percent tin and 40 percent lead with flux at its core. This *flux* in the solder helps to clean the items you're putting together as you solder. We suggest you use a 0.032" diameter solder, which is small enough to help you keep the solder where you want it to go.

✔ **Tip cleaner paste:** This paste is an option for keeping your soldering iron tip neat. Although using a damp sponge (see its earlier bullet) will keep the tip clean for a while, a good cleaning with tip cleaner paste now and then is a good idea.

You can read about soldering safety in Chapter 2.

Drills that come in handy

You will use drills for all kinds of tasks, from attaching wheels to the body of an electronically controlled kart to drilling holes in boxes to fit switches, lights, and much more.

Drills commonly come in ⅜" or ½" chuck sizes. (The *chuck* is the opening in the drill where you insert the drill bit.) This measurement tells you how large of a drill bit (its diameter) will fit in the chuck. For the projects in this book, a ⅜" drill is just fine. Drills come in cordless versions as well as the type you plug into a wall outlet. We prefer cordless drills such as the one you can see in Figure 3-2, along with an assortment of drill bits.

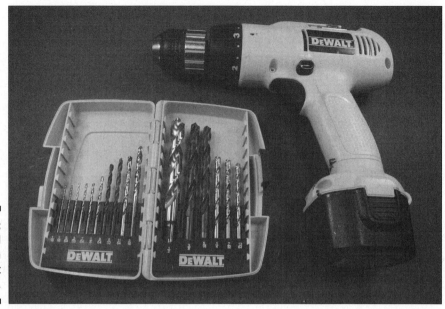

Figure 3-2:
A ⅜" drill
and an
assortment
of drill bits.

For the projects in this book, an assortment of drill bits that includes bits up to ½" in diameter (the shank of the ½" bit should be ⅜") and a cordless ⅜" drill are probably your best bet.

Hacking away with saws

From the magician who saws his assistant in half to the saw you use to cut off a dead tree branch, these tools are handy to trim off excess bits.

Here are the kinds of saws you might need when building electronics projects, especially to build the boxes or boards that contain the electronics brains or trim off little bits of plastic. See Figure 3-3 to view an assortment of saws:

- ✔ A **coping saw** allows you to cut openings in a sheet or box of plastic or wood. For example, you might use this saw to cut a hole in a box to insert a speaker.

- ✔ A **hacksaw** or a conventional **hand saw** can be used to make straight cuts in sheets of plastic or wood or to cut plastic pipes or wooden boards to the desired length.

- ✔ A **mini hacksaw** is useful when you don't have enough room to work with a full-sized saw. This can be common with electronics projects.

Hacksaw

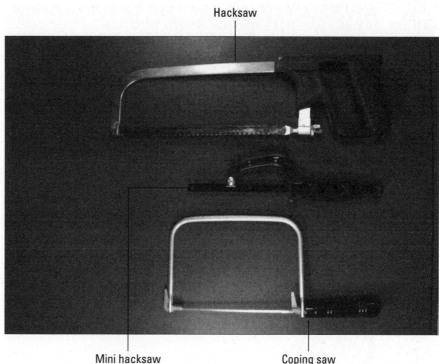

Figure 3-3:
Various
hand saws.

Mini hacksaw   Coping saw

Hacksaws and coping saws have replaceable blades; you can find new blades at hardware stores. When you buy a new blade, check the package to see what material it is intended to cut. Manufactures make blades with different *pitches* (that is, spacing of the teeth) for cutting different materials, such as metal versus wood or plastic.

Power saws go by names such as circular saws, chop saws, jigsaws, table saws, and band saws. You don't need to go out and buy this type of equipment for the type of projects included in this book. However, if you're dying to work with one of these higher-powered saws and someone offers to loan you one, please make sure you know how to operate it safely. It's best to have all your fingers and toes accounted for to work on electronics!

Garden variety tools: Pliers, screwdrivers, wire strippers, and more

This is the category of tools that you're most likely to have floating around your garage or household toolkit. Take an inventory of your toolkit. (We'll wait.) If you're missing any of the tools in this list, it's probably worth going to your hardware store and picking them up.

- **Precision screwdrivers:** This includes both straight and *Phillips head* (the one with the cross shape at the tip).

- **Mini or hobby needlenose pliers:** These are useful for bending wires to various shapes for breadboarding; you also use them to insert wires and components in the holes of the boards.

- **Standard sized needlenose pliers:** These are useful for tasks where you need to apply more strength than mini needlenose pliers can handle. You can see both standard-sized needlenose pliers and the smaller version in Figure 3-4.

- **A small pair of wire cutters:** These are useful for clipping wires in close quarters, such as above a solder joint. The standard size of wire clippers you find at hardware stores is so large that you might have trouble clipping the wire with enough precision. You can see the smaller version in Figure 3-4.

Wire strippers   Precision screwdriver

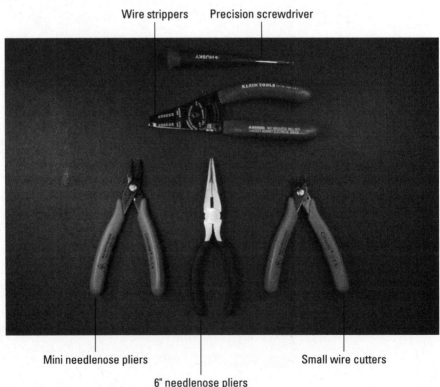

Figure 3-4:
Handy hand
tools.

Mini needlenose pliers   Small wire cutters

6" needlenose pliers

Although you should be able to find small wire cutters and small needlenose pliers at electronics stores such as RadioShack, you might also check out the tools available at hobby supply stores or jewelry-making/bead supply stores. Our local bead supply store carries a nice assortment of tools that work perfectly with small wire.

✔ **Wire strippers:** You use these to cut plastic insulation from the outside of a wire without harming the copper wire inside. The stripped wire can then be inserted into a breadboard or get soldered to a component to keep electricity flowing.

✔ **A vise:** Use this to hold components still while you drill, saw, sand, or whatever.

✔ **A 3X magnifying glass:** This helps you read part numbers on components and check your soldering joints to make sure they're good. You can get handheld or table-mounted magnifying glasses.

✔ **Safety glasses:** Your eyes are one of your most important tools, so be sure to have a pair of safety glasses on hand to protect them. When using the tools in your workshop to drill, saw, clip wires, solder, and perform many other tasks, you need these special glasses to avoid injury from small pieces that could go flying.

Multimeter

A *multimeter* is essentially an electronics troubleshooting tool that you can't do without. You could use it to hunt down the defective part of a circuit — for example, where the voltage is too low to get your circuit going. A multimeter is a combo type of testing tool in that it combines the functions of a few others meters (a voltmeter, an ammeter, and an ohmmeter) in one package.

By using a multimeter, you can take certain electrical measurements, such as

✔ **Current:** The flow of electrons through your circuit

✔ **Voltage:** The force your battery uses to push the electrons through your circuit

✔ **Resistance:** The amount of fight your circuit puts up when voltage pushes the electrons through your circuit

To use a multimeter to test these various measurements, you set a multi-position switch on the meter to have it measure the appropriate range of volts, amperage, or resistance.

Check out Chapter 4, where we tell you exactly how to use a multimeter for more about the various test types it offers and how it works.

Analog or digital multimeters?

Multimeters come in two main types: analog and digital. Think of the difference between a wristwatch that has hands that go around and one that has a numerical readout. For our money (and yours), a digital multimeter is the way to go because you have a smaller chance of making an error when reading the result; even the cheapest is just fine for testing simple projects.

All digital meters have a battery that powers the display. Because they use virtually no power from the circuit you're testing, they're not likely to affect the results.

Auto-ranging is another handy feature to look for in your multimeter if you're willing to spend a little extra money. This sets the test range (see more about this in Chapter 4) automatically.

Test leads that typically come with multimeters use simple cone-shaped tips. You can buy test clips that slip onto the cone-shaped tips to make it easier to clip them onto the leads of a component. This makes testing much easier, trust us.

Components Primer

You use many of the tools discussed in the previous section to work with often teeny, tiny parts called *components.* These range from electrical doohickeys such as resistors and transistors to integrated circuits (chips), switches, and sensors.

Our projects tell you exactly the type and value of component to use (as do most published projects in books and on the Web), so don't worry too much about calculating these values. If you decide to strike off on your own to build your own circuits, we recommend that you get another book, such as *Electronics For Dummies,* by Gordon McComb and Earl Boysen (Wiley) to find out all about understanding component values.

One important terminology point to make is that the terms *pin* and *lead* are almost always interchangeable. They essentially refer to a wire or stamped-out metal bar coming off a component used to connect it to a breadboard or other types of circuit boards. The only exception is *pinout,* which refers to the function of each lead; you never refer to that as a *lead-out!*

Running down discrete components: Resistors, capacitors, and transistors

Whereas *discreet* components should be good at keeping secrets, *discrete* components are so-named because they are one single, solitary thing, rather than a collection of components like those contained in an integrated chip (IC). (Read more about ICs later in this chapter.) Discrete components are single electrical items, typically resistors, capacitors, and transistors.

Diodes are also discrete components, but the only kind we use in this book are light emitting diodes (LEDs). See the later section, "Let there be light: Light emitting diodes," for more about these.

Resistors

The job of a resistor is to restrict the flow of current, which is essentially the flow of electrons. The more electrons, the higher the current. For example, you might want to stop LEDs (which love to eat current) from burning themselves out, and so you would add a resistor to your circuit. You'll find these little guys used all the time in electronics circuits. We measure resistance in ohms, which are so tiny that the measurement of them is usually given in thousands (kohms) or millions (megohms) of ohms. The value of a resistor is indicated by colored bands, with each band representing a number. However, rather than counting colored bands, just read the packaging your resistor comes in or test it with a multimeter.

One variation on a resistor is a variable resistor, also called a *potentiometer*. A variable resistor allows you to constantly adjust from 0 (zero) ohms to a maximum value. These can be mounted on the face of a gadget, where you adjust them with a knob; or, you can mount them on a circuit board, where you have to adjust them by using a screwdriver. A typical use of a potentiometer is to control the volume of an amplifier in a sound circuit.

Capacitors

A *capacitor* stores an electric charge. Quite often, you'll see capacitors used hand in hand with a resistor — for example, in a circuit whose job is to set timing. Because it takes time for a capacitor to fill with a charge, you can set your watch by them (so to speak) if you use a resistor to control how fast the charge (that is, current) flows in. Also, they make good filters for DC signals because although AC passes through a capacitor with ease, DC signals are stopped in their tracks.

Capacitance is a measurement of a capacitor's ability to store a charge. The larger the capacitance, the more charge is stored. You measure capacitance in *farads (F)*. An F is pretty darn big, so you have to use prefixes to show lesser values. The prefixes used are micro- (millionth), nano- (thousand-millionth), and pico- (the ever-popular million-millionth).

Although you can find several kinds of capacitors — based on what material they are made of — three common types of capacitors you'll run across in electronics projects are electrolytic, tantalum bead, and ceramic. Here's are the basic characteristics of each:

- ✔ **Electrolytic capacitors** are typically made of some kind of foil material, and you'll find them with values of 1 microfarad and up. The two types are

    • *Axial:* These have leads stuck on both ends.

    • *Radial:* These have all the leads attached to the same end.

    We use the radial type for the projects in this book because they take less room on a breadboard. The value of this type of capacitor is printed on it along with a voltage rating and its capacitance.

    Be careful to check the voltage rating required by your project and choose a capacitor accordingly.

- ✔ **Tantalum** (a metallic material) **bead capacitors** are available with values of 0.1 microfarad and higher. They cost more than the electrolytic capacitors but are useful if you have a circuit that requires more accuracy because tantalum capacitors have less variation in value than electrolytic capacitors.

- ✔ **Ceramic capacitors** are nonpolarized (see the sidebar, "Polarized counts"), and you can find them with values from 1 picofarad to 0.47 microfarad. Reading the value on these tiny components can be difficult. And to add to the confusion, because many of these capacitors are too small to write the value on in words and numbers, folks use a code. Table 3-1 helps you spot common capacitor values based on their markings.

Polarized counts

Most electrolytic and tantalum capacitors are polarized, so you will see a polarity symbol on them. Typically, only one end is marked with either a plus or minus sign, so you can conclude that the other end is the opposite. With both types of capacitor, the longer lead is the positive one, which is probably the easiest way to identify it.

What's important about being polarized? If a capacitor is polarized, you have to be absolutely sure to install it the right way around in your circuit. If you don't, you will be left with one dead capacitor and possible damage to other components in the circuit.

Small-value capacitors, typically made of ceramic or mica, are nonpolarized so you can connect them any way you want.

Table 3-1	Capacitor Values
Marking	*Value*
101	0.0001 µF
102	0.001 µF
103	0.01 µF
471	0.00047 µF
472	0.0047 µF
473	0.047 µF
474	0.47 µF

For much more detail about various types of capacitors and how to read capacitor values, we recommend that you go out and buy *Electronics For Dummies,* by Gordon McComb and our own Earl Boysen (Wiley).

A final capacitor distinction that we have to make is variable versus fixed. All the capacitors we talked about so far are *fixed,* meaning they have a set value you can't adjust. However, variable capacitors can by adjusted by various methods. We use a variable capacitor, for example, in Chapter 8 for tuning a radio.

Transistors

Transistors are the darlings of the electronics world. Transistors amplify a signal or voltage, or switch voltage on or off. The really amazing thing about transistors is how tiny they are: Before the advent of transistors, people used vacuum tubes to perform the same function, and a vacuum tube is huge in comparison. Transistors also use a lot less power.

When you shop for transistors, be sure to check the package type. You can use packages starting with TO, such as TO-92, TO-39, and TO-220 (as shown in Figure 3-5) in breadboards or manually soldered circuit boards. Packages starting with SOT or SOIC are meant to be used on huge assembly lines and don't have the types of leads that you can use easily in hobbyist electronics projects.

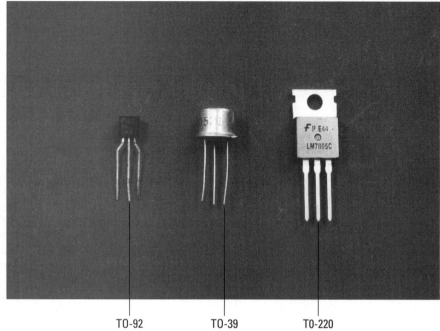

Figure 3-5:
Common
transistor
packages.

TO-92   TO-39   TO-220

Transistors come in

 ✔ **NPN (negative/positive/negative):** You turn on NPN transistors by applying positive voltage; they start to turn on when you apply about 0.7 volts.

   We use NPN transistors throughout this book because it's more straight-
   forward to apply a positive voltage to get things working.

 ✔ **PNP (positive/negative/positive):** You turn off PNP transistors by applying positive voltage; they turn on when you apply negative voltages or
   voltages near ground.

Transistors have three leads: the emitter, base, and collector. In Chapter 4,
we show you how to read schematics so you can figure out where to connect
each pin. For each transistor you use, check the datasheet (which contains a
drawing, called a *pinout*) to determine which pin is which.

ICs

Integrated circuits — commonly known as *ICs* — are like social directors
for components: They gather lots of other components in a single location

(shuffleboard optional). ICs typically contain a number of transistors, resistors, and capacitors connected on a silicon chip to make a functional circuit in one small package.

ICs, as well as some other electrical components, can be susceptible to electrostatic discharge (in other words, *zapping*). For that reason, be sure to also get yourself an anti-static wrist strap (as we discuss in Chapter 2) for your electronics workshop.

ICs come in many packages

Manufacturers make ICs in many types of packages or containers. (A whole valley in California is dedicated to this type of thing.) The type of package that you use either in a breadboard or a circuit board is a *dual inline package* (DIP). A DIP is made up of plastic that encapsulates a silicon chip, with a row of metal leads running on either side of the plastic. You insert these leads into the contact holes in a breadboard and connect components on the breadboard with the circuitry on the silicon chip. (See the later section, "Breadboard Basics," for more about this process.)

DIP ICs are identified by the number of leads they have, such as 8-pin DIP, 14-pin DIP, 16-pin DIP, 18-pin DIP, and so on. Figure 3-6 shows a few common DIP ICs.

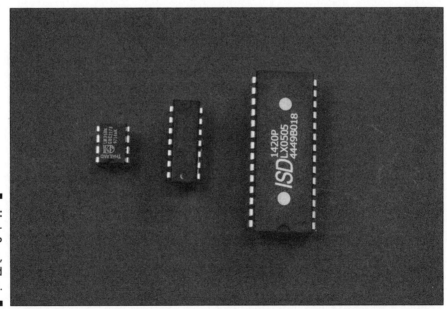

Figure 3-6:
An assortment of DIP ICs (8-pin, 14-pin, and 28-pin).

When you order ICs, be careful to order an IC in a DIP package. If you obtain an IC in a non-DIP package, such as an SOIC (see the earlier section, "Transistors"), you can try forever, but you will never be able to insert the leads into a board. High-volume manufacturers use these other ICs with a technique called *surface mounting*. In this technique, the leads are soldered onto a contact on the surface of a circuit board, not inserted into a hole. If you really need to use an IC that comes only in a surface mount-type package, look for adapters to which you can solder the surface-mount IC and then insert the adapter into your board.

Because ICs are simply a collection of components (such as transistors, resistors, and capacitors) stuffed in miniature onto a silicon chip, each type of IC can perform a different function. That function depends on how many of each component is placed on the chip and how they are wired together. The next two sections provide an overview of two common ICs: operational amplifiers and audio amplifiers.

Opting for op amps

Operational amplifiers (affectionately known as *op amps*) are a type of IC that contains a series of transistors; each transistor amplifies the voltage of the signal just a bit more. You could build a multistage transistor amplifier that could do a similar job by using several transistors, capacitors, and resistors, but why bother? This setup would use about 50 times more space on your breadboard than a single 8-pin DIP op amp.

If you look in a catalog for op amps, you'll probably see pages and pages of them — they're that popular. The fact that we're using 6 volt batteries to power our circuits narrows down the list considerably. Many op amps are designed to work with a positive supply voltage and a negative supply voltage, such as +6 volts and –6 volts. We use op amps in our projects that are designed to work with a 6 volt, or less, single-sided supply. An op amp that is designed to work with a single-sided supply needs only a positive supply voltage and ground.

Op amp ICs usually come in 8-pin DIPs; some of these have one op amp circuit, and some have two op amps *(dual op amps)*. In a dual op amp, the pins that give access to one op amp are on the left side of the DIP, and the pins that give access to the other op amp are on the right side of the DIP. As you can see in the project in Chapter 7, this allows you to build one portion of your project circuit along the left side of the breadboard and a second portion along the right side of the breadboard, which can come in handy if things get overcrowded. Op amp ICs also come in 14-pin DIPs. These contain four separate op amps and are therefore called *quad op amps*.

Here are some common op amps used in electronics projects using low-voltage batteries:

   ✔ **LM358:** A dual op amp
   ✔ **LM324:** A quad op amp

- **MC33171:** A single op amp
- **MC33172:** A dual op amp
- **MC33174:** A quad op amp

Amplifying sound

Audio amplifiers are similar to op amps except that they are designed to provide more power; logically, being audio amps, they provide enough power to drive speakers.

The LM386 is a widely used audio amplifier. Different versions of the LM386 are designed to work with different supply voltages. For example, the LM386N-1, which we use in projects in Chapters 6 and 7, is designed to work with a 6 volt supply and can work with a supply voltage as low as 4 volts. The MC34119 is an audio amplifier that can work with a supply voltage as low as 2 volts.

ICs are available for many specialized uses; you can see some of them in action in the projects that we include in this book. The projects in Chapters 11 and 15, for example, use encoder/decoders ICs paired with motor driver ICs. The project in Chapter 10 uses a timer IC paired with a decade counter IC, and the project in Chapter 7 uses a speech synthesizer IC paired with an audio amplifier IC. The projects in Chapters 9 and 14 both use a voice recorder IC.

The switch is on

A switch seems simple enough: You flick it one way to go on and the other way to go off. However, understanding what's happening behind that switch requires that we give you a bit of background.

- **Open:** A switch is in an *open* state when there is no electrical connection. When switch is open, there is very high resistance between a wire coming into a switch and the wire going out of the switch.
- **Closed:** A switch is *closed* when there is an electrical connection. When a switch is closed, there is very low resistance between a wire coming into a switch and the wire going out of the switch.

There are different kinds of switches, referred to as SPST, SPDT, and DPDT, as shown in Figure 3-7. Here's what these catchy acronyms mean:

- **SPST (single-pole, single-throw):** This kind of switch has two lugs to which you can solder wires. When the switch is on, the two wires are connected; when the switch is off, the two wires are disconnected. We like SPST switches so much that we use them as on/off switches in every project in this book.
- **SPDT (single-pole, double-throw):** This kind of switch has three lugs to which you can solder wires: one for an incoming wire and two for

outgoing wires. When the switch is in one position, the incoming wire is connected to the first of the outgoing wires. When the switch is in the other position, the incoming wire is connected to the second of the out-going wires. (If you have a different need and this is the type of switch you happen to have in your parts bin, you can use just two lugs to make it work as an SPST.)

✔ **DPDT (double-pole, double-throw):** This kind of switch has six lugs to which you can solder wires. These lugs can be attached to two incoming wires and four outgoing wires. When you flip this switch, you simply switch each incoming wire between two of the outgoing wires. We use this type of switch in a relay in Chapter 13 to switch control of the motors from one type of sensor to another.

Figure 3-7:
Three types
of switches
from left to
right: SPST,
SPDT, and
DPDT.

As if switches didn't have enough names, they are also referred to by the method used to change their state from open to closed. See Figure 3-8 to see the different types.

✔ **Toggle switch:** This switch gets its name from the fact that you flip a lever to turn it on and flip it back to turn it off.

✔ **Pushbutton on/off switch:** Every time you push this button, it changes from on to off or vice versa.

- ✓ **Momentary pushbutton switch:** Pushing this switch is what changes its state, but only for the moment! These are also classified by whether they are normally open *(NO)* or normally closed *(NC)*. For example, a momentary normally open switch is closed only while you hold the pushbutton down. When you release the button, it goes back to its normal — open — state.

- ✓ **Tactile switch:** This is a type of momentary pushbutton switch. Tactile switches are rated by the amount of force that is needed to push the button and are often flat so that they can be easily inserted somewhere without protruding (like how we insert them into the hands of a puppet in Chapter 7).

- ✓ **Slide switch:** Logically, this switch operates when you slide a knob to change it from on to off or vice versa.

- ✓ **Relays:** These switches are operated by a voltage rather than by pushing a switch. This makes them very useful for turning on or off a component, such as a light or motor, through a remote control or by voltage generated by a sensor. We control relays with both methods in Chapter 13.

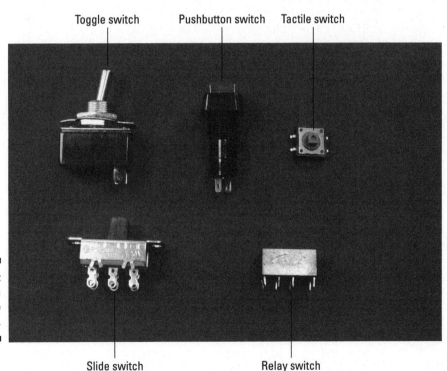

Figure 3-8:
A plethora
of ways to
flip a switch.

Sensors

Sensors take energy in forms such as sound or light and transform that energy into a signal. By using a sensor, you can detect heat, light, and sound, for example. When a signal is sensed, to the sensor produces an electrical signal that is used by your circuit to control some activity. For example, an infrared detector can work in conjunction with an infrared remote control device to stop or start a little go-kart.

Here are a few types of sensors that we use in the projects in this book:

- ✔ **IR detector:** This converts infrared (IR) light into an electric signal. The version that we use in Chapters 11 and 9 contains a photodiode that detects infrared light and an integrated circuit that produces either +V or 0 volts on its output pin. In order to reduce noise from ambient IR light, this detector is designed to only respond to IR light that is pulsed at 38 kHz.

- ✔ **Tilt/vibration sensor:** This type of sensor (which we use in Chapter 14) detects motion or vibrations when the switch is mounted with the body of the sensor horizontal. When the sensor detects motion, it closes a switch, just like a toggle switch works.

Microphones

Technically speaking, a microphone is a kind of sensor. However, there's a lot to say about these sound-sensing devices, so we give microphones their own section (because we're the authors, and we can!).

How condenser (capacitor) microphones work

Capacitors are kind of like a voltage sandwich in that they have two plates, with a slab of voltage between them. A so-called *condenser mike* (also called a *capacitor microphone*) contains one plate made of a very light material that acts as a diaphragm. This plate vibrates when sound waves hit it. This moves the two plates apart, which changes *capacitance* (the ability to store electrons). Moving the plates farther apart decreases capacitance (discharging current), and moving them together increases capacitance (charging current).

Condenser microphones aren't cheap, but they give high-quality sound, so they are often the best choice for an audio-intensive project.

A better mousetrap: Electret capacitor microphones

Today, the most popular type of condenser microphone is the *electret* microphone (which gets its name from the combination of *electro*static and magn*et*), invented in 1962. The electret material used in this type of microphone is made by embedding a permanent charge in a material called a *dielectric*. A charge is embedded in a dielectric by aligning the charges in the material — sort of like how you make a magnet by aligning the atoms in a piece of iron.

There is a preamplifier in an electret microphone, to which you provide a supply voltage. That's why the projects in this book that use electret microphones have a connection through a resistor running between the plus (+) lead of the microphone and the +V bus to power the preamplifier. (The resistor reduces the voltage at the + lead of the microphone to the desired supply voltage.)

Size counts

When you order electret microphones, pay attention to the diameter and thickness because some can be hard to handle and solder. For most of our projects in this book, we use microphones with a diameter of about ⅜" and a thickness of about ³⁄₁₀". A microphone cartridge with a diameter of about ¼" and a thickness of about ¹⁄₁₀" turns out to be much harder to handle and solder to than a microphone cartridge of about ⅜" and a thickness of about ³⁄₁₀". (Check out Chapter 6, where we bit the bullet and used a small microphone cartridge because that project needed some of the capabilities we couldn't find in a larger microphone cartridge.)

Many microphone cartridge sizes are specified in millimeters. To help you translate this, typical diameters of microphone casings are 6 mm (about ¼") and 9.7 mm (about ⅜").

Measuring sensitivity

Sensitivity is another issue that you should pay attention to with microphone cartridges. Sensitivity is measured in decibels (dB) — and just to confuse you, this measurement is given as a negative number. A microphone cartridge with a sensitivity of –40 dB, for example, is more *sensitive* (provides higher voltage at a given level of sound) than a microphone cartridge with a sensitivity of –60.

For example, for the project in Chapter 6 (which has to pick up very faint sounds as part of a parabolic microphone), you need a highly sensitive microphone cartridge. We use one with a sensitivity of –35 dB. In Chapter 14, in which you talk directly into the microphone to record a message, we use a less-sensitive microphone cartridge, rated at –64 dB.

Connecting your microphone cartridge to your project

To connect electret microphone cartridges to your project, you can get electret microphone cartridges with solder pads or with leads that you can insert into a breadboard. We use both in our projects.

Let there be light: Light emitting diodes

A *diode* sends out light when you pass an electric current through it. LEDs, which we use quite a bit in the projects in this book, are similar to the tiny, twinkly lights you use to decorate a Christmas tree, and they come in a variety of colors, such as red, orange, yellow, green, blue, and white. Blue and white LEDs are a lot more expensive, so you don't see them used that often in this book. (We're thrifty!)

LED color isn't controlled by the plastic that surrounds the light. Rather, the semiconductor material used in the LED determines the color. The plastic surrounding the semiconductor material can be clear or treated so that it diffuses the light.

In addition, you can get LEDs in several sizes and shapes. The standard LED, which is a cylinder with a diameter of 5mm, is referred to as *T-1 ¾*.

If you don't connect LEDs the right way, you could wait forever to see the light. Connect the longer of the two leads to the positive voltage and the shorter of the two leads to ground or the more negative voltage.

Speaking up about speakers

Everybody knows what a speaker is: There's one on your DVD player, your computer, your iPod — you name it! Most speakers contain a permanent magnet, an electromagnet, and a cone-shaped device from which the sounds emerge (see Figure 3-9).

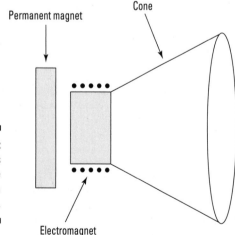

Figure 3-9:
The parts that make up a speaker.

When current moves through the electromagnet, which is attached to the cone, it gets pushed toward or pulled away from the permanent magnet. This depends on which way the electric current is moving. This movement of the electromagnet is what makes the cone vibrate, and that produces sound waves.

Speakers come with a rated *impedance* (the degree to which a component resists electrical current): for example, 4 ohm, 8 ohm, 16 ohm, or 32 ohm. A speaker is often referred to by its impedance: for example, "I'm going out to buy an 8 ohm speaker." When you use a speaker in a circuit, it should have an impedance rating that matches the minimum impedance rating that the amplifier hooked up to the speaker can drive. If you use a speaker with higher impedance than the amplifier can drive, you won't get the maximum amount of sound; conversely, if you use a speaker with lower impedance than the amplifier can drive, you might overheat the amplifier. You can find this rating in the datasheet on your supplier's Web site.

For example, in Chapter 8, we use an 8 ohm speaker because the LM386 amplifier can drive a speaker with impedance as low as 8 ohms. And in Chapter 14, we use a 16 ohm speaker because the ISD1110 voice record/ playback chip can drive a speaker with impedance as low as 16 ohms.

Speakers also come with a power rating, such as 0.2 watt, 1 watt, or 2 watt. Choose a speaker with at least as high of a power rating as the maximum output of the amplifier. Again, you can find this maximum output in the amplifier datasheet.

When you buy a speaker for electronics projects, buy one that comes with convenient holes in the corners of metal or plastic flanges that you can slip screws through. These help you to easily attach the speaker to the box you're putting the circuit in. See Chapter 4 for more about building and assembling projects.

Buzzers

If you have an annoying friend who plays practical jokes, you've probably been on the receiving end of the buzzer and handshake joke. A *buzzer* essentially generates a sound, which you can use in projects in a variety of ways. For example, a buzzer could be the horn on a remote controlled car or an alarm that goes off when a sensor detects motion.

In a buzzer, you apply voltage to a crystal (a *piezoelectric* crystal), which then expands or contracts. By attaching a diaphragm to the crystal, you cause the diaphragm to vibrate when you change the voltage. This vibration causes that *bzzz* sound. There are electromagnet buzzers, but the piezo buzzer works just fine for electronic projects, so we stick to using them in this book.

Most buzzers give off sound in the 2–4 kHz range. Buzzers aren't very discriminating when it comes to voltage: Their ratings are approximate, meaning that a 12V buzzer is absolutely happy to work with a 9V power supply.

Buzzers have two leads, and you have to connect a buzzers the right way round. The red lead is always positive (+).

The Nuts and Bolts of Building Materials

A pure electronics project might just consist of a breadboard containing components and wires. In most cases, though, you'll also want to create some sort of container or chassis to hold the project. For example, if you build a simple radio, you might put the guts of it in a box and drill holes to place the dials and speaker.

You can buy ready-made boxes or other containers and make them work for your project. You can also build your own out of various materials.

Plastic

ABS (please don't ask what this acronym means; we could tell you, but you couldn't pronounce it) plastic boxes are available from most electronics suppliers. These are lightweight, sturdy, waterproof, and handy for housing your gadgets. We use a plastic box in Chapter 11 to house a remote control.

The plus with plastic boxes is that components such as switches are designed to be mounted on boxes or panels with thin walls. Therefore, mounting these components on these plastic boxes is often easier than on wooden boxes.

The downside is that cutting clean openings in plastic is harder than in wood — for example, to insert speakers.

Wood

We like to use wooden boxes to house many of our projects because, well, they look nice. We found simple, unfinished wooden boxes at a national craft supply superstore (Michaels), but any craft store probably stocks them.

Wood is also easier to drill and cut than plastic; however, you'll often find that the wall of the box is ¼" thick, which makes mounting components such as switches more complicated. In Chapter 4, we provide some advice on how to handle mounting components on wood.

Build it yourself

If you don't like to buy ready-made containers, you can make your own boxes from wood or plastic. You can easily find lots of books that tell you how to make all sorts of things from wood, so we don't get into that. To start, you might check out the Woodbox.com Web site and click the Wooden Box Making link for a good overview.

If you want to make your own boxes or build fancier shapes out of plastic, such as a model car, to mount your electronics, check out

www.talkingelectronics.com/Projects/Boxes/BJones-BoxArticle.html

This article is a nice introduction to making custom plastic boxes for electronics projects.

For some projects, you will mount boxes on a base or sandwich them between two sheets of materials. We used sheets of PVC and plywood in our projects. Quarter-inch or 6 mm thickness is a good bet for a strong base. You can use thinner sheets — for example, ⅛" or 3 mm — when you don't need structural strength. Rigid, expanded PVC is often used instead of other plastics because it resists the buildup of electric charges, which might cause electrostatic discharge, which can zap your electronics components.

Robot supply houses, such as Solarbotics (www.solarbotics.com) or Budget Robotics (www.budgetrobotics.com), carry PVC sheeting. At Solarbotics, this material is sold under the product name Sintra. You can find plastics suppliers that sell rigid expanded PVC in 4' x 8' sheets, which is economical if you plan to use a lot of it.

Holding it all together

Sticking materials together to form boxes or whatever can be done in a few different ways.

You can attach many different types of materials together with glue. Look for a glue called *contact cement.* This can bind a wealth of materials, including metal, plastic, rubber, and wood.

To mount components such as speakers, you need screws and nuts. The parts lists in our project chapters tell you what size screws and nuts to get; we're betting you have some of these in that leftover cake tin in your garage gathering dust, but you can buy what you need for pennies in any hardware store.

We find that 6-32 screws fit many mounting holes.

Holding down wires

Wire clips are very useful for organizing wires that you affix to your project container. These generally have an adhesive backing on the base that you use to attach them to a surface. Then you slip the wires into the clip, and they are nicely held in place. (We use RadioShack part #287-1668.)

Cable ties can also be useful when you want to run wires along something without a flat surface, like a wooden dowel.

Breadboard Basics

A *breadboard* is a rectangular plastic box filled with holes, which have contacts in which you can insert electronic components and wires. A breadboard is what you use to string together a temporary version of your circuit. You don't have to solder wires or anything else; instead, you poke your components and wires into the little contact holes arranged in rows and connected by lines of metal; then you can connect your components together with wires to form your circuit.

The nice thing about breadboards is that you can change your mind and replace or rearrange components as you like. You typically create an electronics project on a breadboard to make sure that everything works. If it's a project you wish to save, you can create a more permanent version. We use breadboarded versions of circuits exclusively in this book.

If you want to create a permanent version of your circuit, you need to create a soldered or printed circuit board; see the sidebar, "Printed circuit boards," to find out how to go about that.

There are a few different sizes of breadboards, some of which are shown in Figure 3-10. You can link breadboards to make a larger circuit, like the one shown in Figure 3-11. See Chapter 4 for more about how to build a breadboard.

Figure 3-10:
Two bread-
boards, one
with 830
contact
holes and
one with
400 contact
holes.

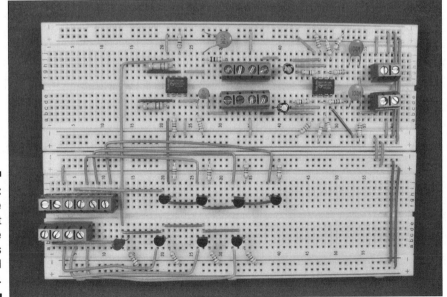

Figure 3-11:
A large
circuit built
on multiple
breadboards
hooked
together.

Printed circuit boards

If you create a circuit on a breadboard and decide that it's worthy of immortality, you can make it permanent by soldering components in place on a printed circuit board. To do this, you have to get your hands on a universal printed circuit board. This is much like a breadboard except that you can solder all the connections you've made to keep them around.

A universal printed circuit board has rows of individual holes throughout the board with copper pads around each hole and metal lines connecting the holes in each row, like in a breadboard. You mount parts on the face of the board and then pass leads through holes to the components. You can solder the leads to the copper pads on the bottom of the board. Universal printed circuit boards are available in a variety of patterns of contact holes and metal lines. The figure here shows one we like because there are rows on either side that accommodate discrete components handily. This circuit board is made by One Pass, Inc. (www.onepassinc.com).

You can get custom printed circuit boards made for your circuit; this is typically done by submitting a drawing of your circuit to a printed circuit board company. These boards eliminate the need to solder jumper wires between components.

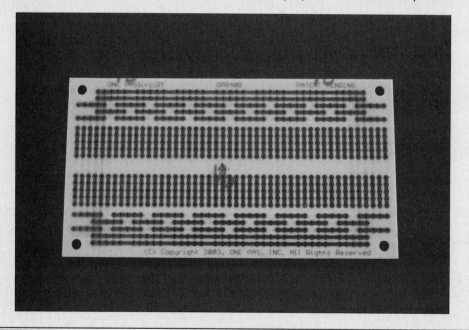

Wires pull it all together

When you place components in a breadboard, you don't get much action until you connect those components with wire. Wire used in electronics is copper surrounded by a plastic insulator, usually called *hookup wire*. Hookup wire comes in various diameters referred to as a *gauge*. The standard gauge measurement used in the U.S. is American Wire Gauge, also referred to as *AWG*. We generally use 22 gauge or 20 gauge wire.

Someone decided at some point that the smaller the gauge, the larger the diameter of wire. For example, 20 gauge wire is 0.032" in diameter, and 22 gauge wire is 0.025" in diameter. Don't ask us why — just memorize this fact!

We use 22 gauge solid wire for most of the projects in this book. (Okay, in two chapters, we use 20 gauge; and in one, we use 26 gauge, but you'll find out why when you get to those projects.)

Use solid wire — never stranded wire — between components within a breadboard because stranded tends to separate when you try to insert it into the holes of a breadboard.

You can buy hookup wire in spools; we generally get spools containing 100 feet of wire. If you are starting with only a few projects, you can get smaller spools containing as little as 30 feet of wire.

The insulating plastic that surrounds wire is made in different colors. Pick up a spool of red and a spool of black. Using different colors helps you to identify the purpose of different wires in your project.

You might also consider buying an assortment of different lengths of pre-stripped and prebent 22 gauge wire jumpers. *Jumper wires* — which are used to connect components in a breadboard — save you a lot of time you might otherwise spend cutting small wires to length, stripping them, and bending the stripped wire when you're building a breadboard.

Insulating those naked wires

You will also use various materials to insulate bare wires. You can use electrical tape, for example, to insulate solder joints that might touch each other and cause a short.

Heat shrink tubing is a tidy way to insulate the point where wires connect in a solder joint. Heat shrink tubing is simply a plastic tube. When you slip a short length of this tubing over a solder joint and apply heat, the plastic tube shrinks, providing an insulating layer around the wire. When working with 22 gauge wire, we use $\frac{3}{32}$" heat shrink tubing.

Liquid electrical tape is also handy to insulate bare wire in situations where heat shrink tubing or conventional electrical tape doesn't work very well. We use liquid electrical tape in Chapters 5 and 10, for example.

Connectors

Finally, terminal blocks are used to connect wires from components such as speakers, motors, and microphones to the breadboard. A *terminal block* is a small block of plastic that you mount on a breadboard. You insert the wires into the terminal block through a hole in the block and then tighten screws to hold the wire securely.

When choosing terminal blocks, the diameter of the pin that inserts into the breadboard or circuit board is important. Some terminal blocks that work fine in circuit boards where the components are soldered in do not stay in place on breadboards. We have had the most success with RadioShack part #276-1388.

Chapter 4

Running Down the Skills You Need

You need three key things to build an electronics project: materials, a workspace, and certain skills that help you assemble the materials into something that moves, beeps, lights up, or otherwise makes itself electronically known to the world. (You also need time, but we'll leave that to you to find.)

In Chapter 3, we talk about organizing your materials and workspace. In this chapter, we address the third element — the skills that you need to read circuit designs, build, and test electronics projects. This chapter provides an overview of these skills; for more comprehensive explanations, check out *Electronics For Dummies,* by Gordon McComb and Earl Boysen (Wiley).

It's Symbolic: Reading a Schematic

A *schematic* is your blueprint for building an electronics project. A blueprint for building a house uses various symbols to represent elements, such as doors, and lines to show walls. Instead of doors and walls, the symbols and lines in a schematic represent components such as transistors, integrated circuits (ICs), and resistors as well as the wires that connect them.

Be sure to check out the Cheat Sheet (the yellow page at the front of this book) for a table of commonly used schematic symbols.

Schematics help you understand how a particular electronics project works as well as how to build it. You can build the circuit on a breadboard (more on that in the upcoming section, "Breadboarding") by inserting the components and making the connections on the board that are indicated by the schematic.

Perusing a simple schematic

An example of a very simple schematic shows a battery, one electronic component, and the wires connecting them. Figure 4-1 shows a schematic that contains a 1.5 volt battery, a wire from the positive side of the battery (+V) connecting it to one of the leads on an LED, and a wire connecting the other lead of the LED to the negative side of the battery. With both wires connected, current flows from one terminal of the battery through the LED, making it light up, and back to the other terminal of the battery. (If the LED is connected to only one battery terminal, no current flows, and the LED will not light up.)

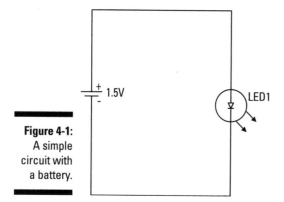

Figure 4-1:
A simple
circuit with
a battery.

Some circuits use too many components for the schematic to show the wire connecting every component to the battery. In those cases, we use a convenient symbol for a voltage source to represent the positive side of the battery and a ground symbol to represent the negative side of the battery, as shown in Figure 4-2. This is same circuit as the preceding figure with a voltage source symbol and ground symbol representing connections to the battery. These symbols are also used in applications where a metal chassis is used as ground and a power supply is used to supply voltage.

You can read more about connecting to +V and ground in the later section, "The anatomy of a breadboard."

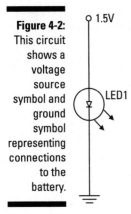

Figure 4-2:
This circuit
shows a
voltage
source
symbol and
ground
symbol
representing
connections
to the
battery.

Interconnections among components (how wires are connected to move electricity from one component to another) in a circuit are made with wires or bits of copper placed on a breadboard. A schematic won't usually specify which kind of connection you are using, only that a connection exists. Figure 4-3 shows a few methods of representing interconnections.

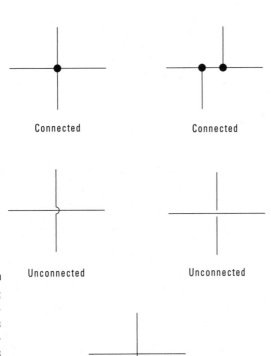

Figure 4-3:
How connections and non-connections are represented in schematics.

You'll also find symbols for commonly used components, such as resistors, diodes, capacitors, and transistors, in schematics (see Figure 4-4). Chapter 3 explains what most of these components do in a circuit.

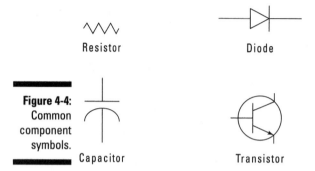

Resistor   Diode

Figure 4-4:
Common
component
symbols.

Capacitor   Transistor

Switching gears with switches

You use *switches* to turn power to a circuit on or off or to connect/disconnect a pin on a component to +V or ground. However, switches don't use just a single symbol. Rather, they come in several varieties, indicating

- How many wires they control
- Whether they stay in the position you set them at or return to a normal position after you release them

An *SPST (single-pole, single-throw) switch* has one incoming wire and one out-going wire that you use to open or close a connection in a circuit. For example, if a wire runs from the negative pole of a battery to an SPST switch and another wire runs from the SPST switch to a circuit, current can flow through the circuit when you have the switch in the closed position. When you flip the switch to the open position, no current can flow through the circuit.

SPST switches also come in the momentary switch variety; these can be nor-mally open (NO) or normally closed (NC) and are generally controlled by pushbuttons or relays. A normally open switch conducts current only when the button is pressed and returns to its open position when it's released. A normally closed switch won't conduct current when you press the button but returns to its normal position and conducts current when you release it.

An *SPDT (single-pole double-throw) switch* has one incoming wire and two out-going wires that you use to control which of two components is connected. Suppose that the incoming wire is connected to power, one outgoing wire is connected to a green LED, and the other outgoing wire is connected to a red LED. When you have the switch in one position, the green LED lights up; when you flip the switch, the green LED goes dark, and the red LED lights up.

You can think of a *DPDT (double-pole double-throw) switch* as containing two SPDT switches that switch in tandem. To see an example of this in action, visit Chapter 13, where we use DPDT relays to simplify the wiring of our breadboard.

Schematic variables

Some components are *polarized,* which means that you have to put them in the circuit in a particular way. Schematics can identify the polarity of components (see Figure 4-5).

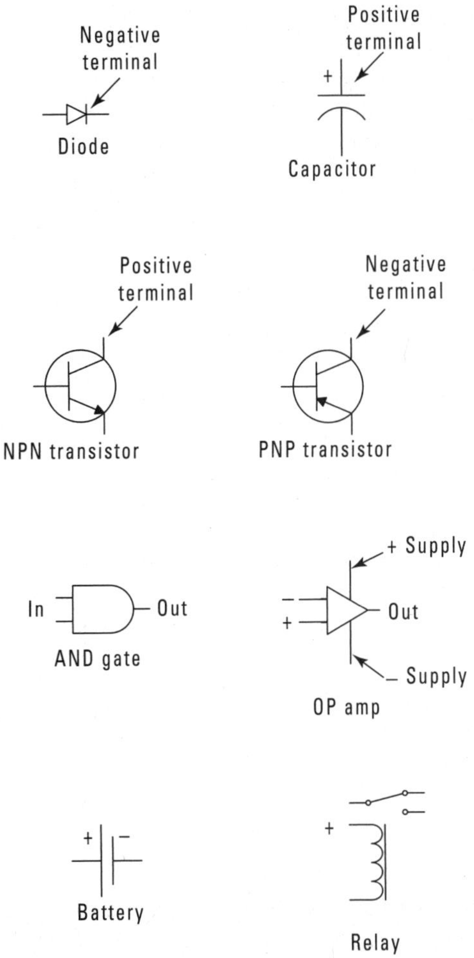

Figure 4-5:
Common
polarity
symbols.

Identifying the + lead on polarized capacitors and LEDs is easy because the + lead is longer than the ground lead. For transistors and integrated circuits (ICs), you have to take a look at the datasheet to find out which pin to connect to +V and which pin to connect to ground. A *datasheet* is the manufacturer's specifications for the component. You can read about pins in Chapter 3.

Some components are *variable*, meaning that they don't just operate at one value; instead, you can adjust their values. Variable resistors (also called *potentiometers*), variable capacitors, and variable coils are all examples of adjustable components. You can use these adjustable items to control volume or tune in a radio station, for example.

Pulling it all together

After you understand some of the elements that go into a schematic, we thought you'd like to have us rundown how to read a simple schematic sample. The schematic used in Chapter 6 is shown in Figure 4-6. This is a circuit that includes a microphone and an IC that amplify noises; the circuit works together with a *parabolic* (curved) metal dish that helps pick up sounds.

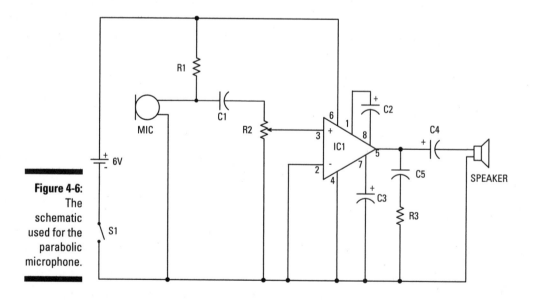

Figure 4-6:
The schematic used for the parabolic microphone.

First, some general rules: A line between two symbols indicates that the two components are connected by a wire. A connection is also indicated when a symbol is shown connected by one line to another line with a dot at the junction.

Here's what this schematic is saying:

- ✔ A **battery** is used to supply 6 volts to the circuit.

- ✔ S1 is an **SPST switch** that turns the power to the circuit on or off.

- ✔ An **electret microphone** (MIC) transforms sound waves into electrical signals.

- ✔ A **resistor** (R1) connects the microphone to the positive battery terminal and supplies the 3 volts required to make the microphone function. Note the dots above and below R1 that indicate connections.

- ✔ C1 is a **capacitor** connected between R1 and R2.

- ✔ R2 is a **potentiometer** with one lead connected to C1, one lead connected to the negative battery terminal, and the variable contact connected to Pin 3 of R2.

- ✔ IC1 is an **audio amplifier (op amp)** connected at Pin 3 to R2.

- ✔ **Pins 2 and 4 of IC1** are connected to the negative battery terminal.

- ✔ **Pin 6 of IC1** is connected to the positive battery terminal.

  Capacitor C2 is connected between Pins 1 and 8 of IC1. The positive side of the capacitor is connected to Pin 1.

- ✔ **Capacitor C3** is connected between Pin 7 of IC1and the negative battery terminal.

- ✔ **Capacitor C4** is connected between Pin 5 of IC1 and the speaker (or headphones).

- ✔ **Capacitor C5** is connected between Pin 5 of IC1 and resistor R3.

- ✔ **Resistor R3** is connected between capacitor C5 and the negative battery terminal.

- ✔ **The speaker (in this case, headphones)** is connected between capacitor C4 and the negative battery terminal.

Breadboarding

A *breadboard* is a temporary place to build and test a circuit for an electronics project. You don't have to solder the circuit; just insert components and the wires, connecting them into handy little holes.

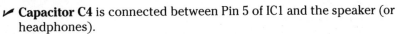

When you're sure you have your circuit right, you can create permanent boards by soldering or by ordering printed circuit boards. See *Electronics For Dummies,* by Gordon McComb and Earl Boysen (Wiley), for some detailed descriptions of these processes.

The anatomy of a breadboard

The breadboard itself is a plastic board with strips of metal running underneath and holes in the top. You slot the little wire legs that sprout from components and also the connecting wires into these holes, which contain metal channels called *contacts*. The metal strips that run underneath connect the items you plug into the holes to each other and the battery.

Breadboards come in various sizes; however, no matter what the size, the top and bottom rows of contacts on the breadboard (see Figure 4-7) are linked horizontally, and you'll typically use them to connect to your battery.

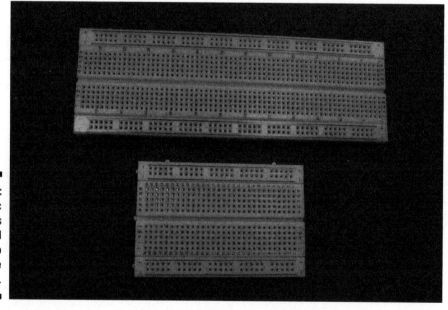

Figure 4-7:
Your basic
solderless
breadboard
in our two
favorite
sizes.

TIP

How to figure out what size breadboard to get? Some have as many as 3,200 contact points! But don't overdo. For the projects in this book, we used boards with 400 contact points for small circuits and 830 contact points for medium circuits; for our large circuits, we hooked two boards together with the handy ridges and notches on the sides of the boards.

Notice the + and – (negative) signs on the breadboard. The positive battery terminal is connected to the rows with the + sign; these rows are often referred to as the *+V bus*. The negative battery terminal is connected to the rows with the – sign; these rows are often referred to as the *ground bus*. Because the +V bus and the ground bus run the entire length of the board on both sides, you need to use only a short piece of wire to reach a +V or ground bus from anywhere on the breadboard.

Other contacts on the breadboard are linked vertically in rows of five; the five points are connected electrically by metal strips. Most folks place chips in the middle of the circuit, straddling the little aisle with each pin of the IC in a contact hole. That way, each pin of the IC is electrically connected to four other contacts, making it easy to connect other components to IC pins.

Don't fry your board! These things are very susceptible to heat. Shorted components can melt them. Check the components with power on to make sure that nothing overheats. Also, they are designed only for low-voltage DC projects, so don't apply too much juice.

Figuring and finessing the layout

How you arrange items on a breadboard won't look exactly like how you've arranged items in a schematic. You have to pay attention to a few issues when laying out components on your breadboard. The schematic shows the elements of a circuit and connections, but a breadboard is arranged to make the most efficient connections possible using the holes and connectors available. Here are some tips to keep in mind.

Pin numbering: ICs have pins that are numbered counterclockwise, starting at a little notch or dot indicator (see Figure 4-8). You should place all ICs pointing in the same direction. This helps you avoid inserting an IC backwards and also helps you keep track of the pin numbers. Use various pins, as specified on the IC datasheet, to connect to +V, ground, and other components.

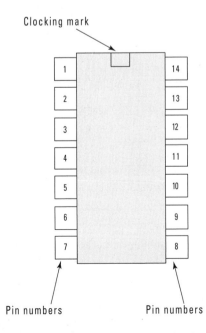

Figure 4-8:
The IC pin numbering schematic.

Neatness counts: Take your time to make your board neat and tidy. This helps you to avoid mistakes and also helps you to troubleshoot if things aren't working quite right.

Spacing: Leave yourself room to place items, allowing a little space between them. It's better to leave a little more space between elements and use a bigger or expanded breadboard than to crowd yourself too much. This gives you the space to modify and refine your circuit.

Jumps: Minimize the jumps that you make between connections. (Typically, this involves using a jumper wire.) For example, if you can insert one lead of a component in the same row as the lead of another component you're connecting to, you don't have to use a jumper wire to connect them. The less wiring you have, the less messy things get.

Using color-coded wiring helps you to keep track of your layout. For example, many people use black wire for ground and red wire for power. Put wires at 90° angles, not on the diagonal, because diagonal wires will get in the way of other components to be placed on the board.

Also keep wires to a practical length: that is, long enough so you can route them around ICs but short enough so you don't have lots of extra wire cluttering up your breadboard. (Routing wires around ICs means that if you have to remove or replace an IC, you don't have to remove all the wires as well!)

Assorted lengths of prestripped wires are available that save you time in cutting, stripping, and bending wires to length. You just pick one that's already cut to the right length. However, each length of these wires is a different color; thus, if you use prestripped wires, you can't color-code your circuit. Because most of the photos in this book are black and white, we went for the convenience of using prestripped wires rather than color-coding the wires on the breadboard.

Inserting wires and components

In a nutshell, here's how to wire a breadboard:

1. **Use 22 gauge solid wire to make connections (see Figure 4-9).**

   Don't use stranded wire because it can get smushed when you push it into a hole and could even cause shorts in your circuit if a piece of wire breaks off.

   See the sidebar, "When stranded wire works," for times when using stranded wires is more appropriate.

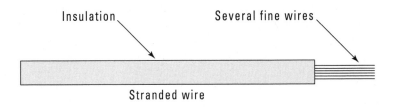

Insulation

Several fine wires

Stranded wire

Figure 4-9:
Stranded
versus solid
wire.

Insulation

Single wire

Solid wire

2. **Measure how long of a wire you need to make each connection.**

3. **Strip off ¼" of insulation from each end.**

   Better yet, buy prestripped wires.

4. **Bend the bare wire at a right angle.**

5. **Insert the wire into a hole in the board.**

The schematic shown earlier in Figure 4-6 is shown translated onto a bread-board in Figure 4-10. You don't see the potentiometer, microphone, battery, switch, or speaker on the breadboard because these are connected through wires attached to the five terminal blocks (TB). The sole purpose of a *terminal block* is to provide a place where you can attach wires to your circuit board by inserting them into holes and using a screw to clamp them down.

Notice that we inserted a lead from C2 into a contact in the same row as Pin 1 of IC1, thereby making electrical contact. We ran a wire from the same row as the other lead of C2 around IC1 to Pin 8 of IC1. This produces a lot neater board than you get when you loop wires over the IC.

Notice also how all the wires are flat on the breadboard. We cut them all to the length required so they didn't have excess wire poking up in the air.

We measured the resistor leads so that we had enough length to cross the distance between the contacts and still have about ¼" more on either side to bend down and insert in the breadboard holes.

C3 C5 R3   C1   Battery and S1 TB

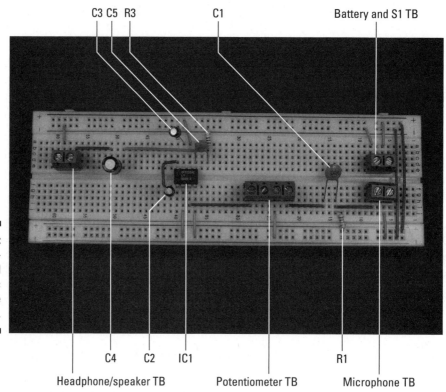

Figure 4-10:
A bread-
boarded
parabolic
microphone
circuit.

C4   C2   IC1   R1

Headphone/speaker TB   Potentiometer TB   Microphone TB

We bent the wires on the ceramic capacitors at a 45° angle so that the face of the capacitor is visible. That way, you can easily read the value of the capacitor on the board.

We cut the leads of the electrolytic capacitors to about ¾" to minimize how far they stick up in the air.

When you use a breadboard, you can use and reuse components for different projects. However, be aware that the little contact wires on components can break off easily. If you remove ICs, use an IC extractor or the flat end of a small screwdriver to pry the IC up at both ends, or you will damage it and probably end up tossing it. The leads on ICs aren't designed to be bent more than once or twice, or they will break off.

Soldering Your Circuit Board

Electronics projects involve a lot of little bits called *components* (think transistors and capacitors, for example) and wire, and items like microphones and light bulbs and such. In many instances, you have to solder some of these things together to provide an electrical connection between them. *Solder* is a metal material that you melt and apply to two items; when it cools, it forms a joint that holds items together and forms an electrical connection.

So why do you need to solder if you use solderless breadboards? Although we chose to not have you make circuits permanent by soldering them for the purpose of the projects in this book, we do ask you to solder wires to switches and microphones and such, so this is definitely something you need to be able to do.

Soldering perfect joints is an acquired skill — one that you just get better at with practice. Here are some valuable tips for getting started.

Please, please read the several safety precautions about soldering in Chapter 2. You're playing around with 700°F temperatures here, and we don't want you to get hurt!

Using a soldering iron

A *soldering iron* (sometimes called a *soldering pencil*) is like a wand that gets very, very hot so that when you touch it to the solder, it melts it. You can find a variety of soldering iron models (see an example in Figure 4-11), which will vary in price based on features, such as those we discuss in Chapter 3.

Figure 4-11: A soldering iron.

When you're ready to solder, make sure you attach the best tip for the job; a smaller conical or chiseled tip is your best bet. Then, make sure that the soldering iron is firmly seated in its holder. Finally, wait for it to reach the right temperature, somewhere around 700°F. Just touch the end of your solder to the tip, if the solder quickly melts, the iron is hot enough.

Before using a new solder iron — and periodically, as you use your iron — you should *tin* it (coat the tip with solder):

1. **Heat up the iron.**

2. **Clean the tip by wiping it on a moist sponge.**

3. **Apply a little bit of solder to the tip.**

4. **Wipe off any extra solder with a moist sponge.**

Working with solder

Solder is a rather soft metal, and the most common type for electronics projects is a 60/40 rosin core. The rosin core contains *flux,* which cleans the surface of the wires being soldered. This helps the solder stick to the wire surface.

Solder also comes in different diameters. You don't need super-thick solder for electronics projects. We use 0.032" diameter solder on the projects in this book.

Molten solder sends out fumes that you wouldn't want your worst enemy to breathe. Lead-free solder helps you avoid toxic lead fumes. Keep your workspace well ventilated no matter what kind of solder you use.

When you solder, you press the cold (solid) solder to a part and then apply heat to a part you want to join, not to the solder itself (see Figure 4-12).

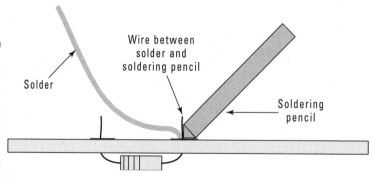

Figure 4-12: In this example, apply the iron to the wire, not the solder.

Solder

Wire between solder and soldering pencil

Soldering pencil

When you solder, hold the soldering iron just as you would a pencil (near the base) and be careful to avoid touching the very hot tip. Touch the iron to the elements that will be joined to heat them and then feed solder onto them. The solder should flow like how water flows around your finger when you hold it under a running faucet.

When you're done soldering, pull the solder and the iron away, and let the solder cool that you applied. Take a look at the joint you made; it should be shiny and shaped like a little mountain (not a deflated soccer ball).

Here are some tips for good soldering:

✓ **Keep it clean.** Make sure the parts that you solder are clean and that your soldering iron has a clean, tinned tip.

See the preceding section for the skinny on tinning.

✓ **Watch the heat.** Be sure to get the soldering iron hot enough and heat any parts you are soldering before you apply the solder.

✓ **Easy does it.** You should need to hold the soldering iron on a joint only a few seconds.

If you heat a component for longer than a few seconds, you might damage it.

✓ **The eyes have it.** Always wear safety glasses when soldering.

Pockets in solder could pop when heat is applied. Your eyes are not the place for hot solder to settle.

✓ **Keep it clean, Part 2.** Keep a damp (not dripping wet) sponge handy to wipe away excess solder on the tip and to wipe the tip clean before soldering each component.

✓ **Bend before you solder.** Before soldering a wire onto a component, bend the end of the wire in a U shape and insert the U through the hole in the lug you want to solder to. Use a pair of needlenose pliers to clamp the wire to the lug. Then you can solder without having to hold the wire, the solder, and the soldering iron, which is nigh impossible (assuming you have only two hands).

Read about third-hand clamps in the upcoming section, "Soldering extras."

Figure 4-13 shows a switch and two potentiometers with wires soldered to the component lugs.

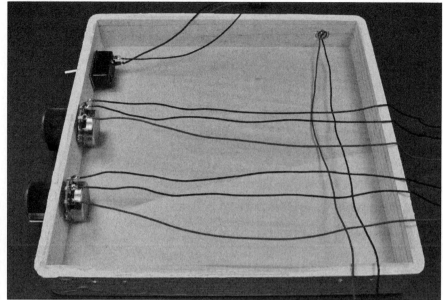

Figure 4-13:
Soldering
wires to
component
lugs.

✔ **Bend before you solder, Part 2.** Before soldering a wire to a component that has presoldered flat contact pads, do the following:

1. *Bend the end of the wire at a 45° angle.*

2. *Heat the end of the wire.*

3. *Apply a light solder coating to the wire.*

4. *Press the wire onto the contact pad with the soldering iron.*

5. *Hold down the wire with the soldering iron until the solder on the pad melts.*

6. *Remove the iron and hold the wire on the pad with your other hand until the solder cools. (You should hold the wire several inches from the solder joint so your fingers don't get hot.)*

Make sure that the component you're soldering is kept steady. (Read about third-hand clamps in the next section for help with this.)

Figure 4-14 shows a microphone cartridge wire soldered with this technique.

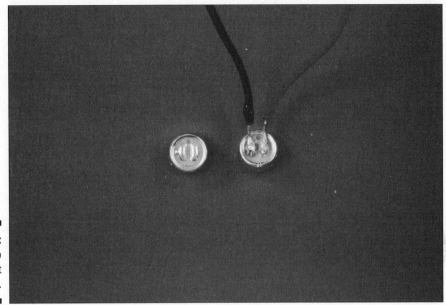

Figure 4-14:
Soldering to
flat contact
pads.

Soldering extras

Your soldering iron and solder are the main tools you need to make soldered joints. However, a few accessories will make your soldering life easier. These include

- ✔ **Sponge:** You use a damp (not dripping) sponge to wipe the tip of your soldering iron clean before soldering each component.

- ✔ **Tip cleaner:** If you don't keep the iron's tip clean, it might actually repel solder — making it bead up and staying away from where you want it to go. When the tip is too grungy to be cleaned by simply wiping it on the damp sponge, use a tip cleaner paste to chemically clean it.

- ✔ **Solder wick:** Sometimes you have to desolder a bad joint and then resolder it. To help remove the bad solder, you can use a *solder wick,* which is a flat, braided piece of copper that soaks up solder.

- ✔ **Third-hand clamp:** There are fancy clamps you can buy called third-hand clamps to hold components while you solder them. Personally, we just use a vice and an alligator clip; they do just fine!

Measuring Stuff with a Multimeter

A *multimeter* (see Figure 4-15) is a testing device that, um, tests multiple things, including resistance, voltage, and current. Using certain multimeter models, you can test to be sure that components — such as diodes, capacitors, and transistors — function properly. You can also troubleshoot your circuit to see where current is failing and pinpoint the problem spot.

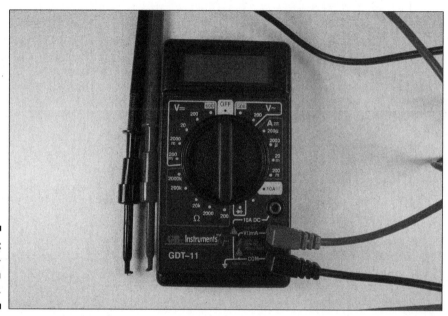

Figure 4-15:
The multi-function multimeter.

You don't have to break into your piggy bank to buy a multimeter. You can find them for about $10; if you want fancy features, you can spend over $100. Find a model whose price you like and then splurge on the next higher-priced model. You will use a multimeter all the time. Trust us: It's worth a few extra bucks for a better model. See Chapter 3 for more information about multimeter features.

How a multimeter works

A multimeter has a set of leads: a black one and a red one. You attach these leads to the component or portion of the circuit that you're testing, and a digital readout provides the results. You adjust a knob to set the test you wish to perform such as resistance, voltage, or current as well as the range to test. *Note:* Some multimeters have an auto-ranging feature that saves you the trouble of setting the range.

Test leads that typically come with multimeters use simple cone-shaped tips. You can buy test clips that slip onto the cone-shaped tips to make it easier to clip them onto the leads of a component. This makes testing much easier, trust us.

The two things we test most often with our multimeter are resistance and voltage.

Reading resistance

The problem with resistors is that manufacturers seem to expect you to memorize the color code that identifies the resistance rating. Here is an easier way:

1. **Clip your test leads onto the resistor leads.**

2. **Dial your multimeter to the resistance range you think the resistor fits in.**

3. **Read the value.**

If your multimeter reads 1, you guessed too low of a value. Move the dial to the next range up until you get a valid reading. If your multimeter reads at close to 0 (zero), you guessed too high of a value. Dial to the next range down until you get a valid reading; if you get to the lowest range and the value is still 0, whatever you're testing has zero resistance.

Testing switches or relays is another common use of the resistance-testing feature of your multimeter. You can clip your test leads onto the lugs of an SPST switch to verify that it's working. (***Hint:*** Occasionally, they don't work.) When the switch is open, you should get a value of 1, meaning that the resistance is higher than your meter can measure. When the switch is closed, you should get a low resistance — close to 0 (zero) ohms. You can also test SPDT or DPDT switches or relays, like those we use in Chapter 13, to make sure which lugs are connected in which switch position.

Measuring voltage

To run a test to measure voltage, you connect the red multimeter lead to the positive side of the battery or circuit that you're testing and the black lead to the negative or ground side and set the dial to the voltage range you expect.

We often check the voltage at the contacts of a battery pack. To do this, touch the red lead to one of the battery pack outputs and the black test lead to the other. With a 4-battery pack loaded with fresh batteries, you should get a reading of about 6 volts. (If you get a reading of –6 volts, don't worry: Just reverse which lead you are touching to which battery pack output.) When batteries get old, the voltage drops. If you get less than 5 volts from a 4-battery pack, it's time to get new batteries.

When a circuit doesn't work, one of the first things to check is the voltage between the +V bus and the ground bus of the breadboard. Here's how:

1. **Strip both ends of a 3" piece of 22 gauge wire.**
2. **Clip one end of each wire to one of your test leads.**
3. **Slip the free end of the wire attached to your red test lead into any contact on the +V bus.**
4. **Slip the free end of the wire attached to your black test lead into any contact on the ground bus.**

   Although you might not get a reading of the full 6 volts because of drain on the battery from the circuit, you should get a reading above 3.5 volts.

   If you get a reading close to 0 (zero) volts, check to make sure that your battery pack and the wires from the battery pack terminal block are connected properly.

Working with the Boxes that Contain Your Projects

In most cases, you'll want to put the breadboard on which you build your circuit into some kind of container. A container can make toting around your breadboard easier, help prevent little bits from falling off, and make your project look better. You might also want to add mechanisms for controlling your circuit in a box. For example, you might operate a remote control device by disconnecting and connecting wires on a breadboard, but wouldn't it be easier to put the breadboard in a box and then add switches and buttons you can use to make it work?

In this section, we give you some advice about basic skills you need to work with these containers for your projects.

Working with boxes

Essentially, using a box involves finding the right type of box and then drilling or cutting holes in it to poke wires and items such as switches or speakers through.

Choosing plastic or wood

You could build your own boxes, but you can find a large variety of containers that you can simply buy cheaply and put right to work, including plastic and wooden boxes in various shapes. We typically put remote control circuits in plastic boxes to make the control light and compact for handling. And we typically use wooden boxes to house other circuits because we can make them look a little more stylish than the plastic boxes.

Drilling and cutting holes

We use a drill to make holes up to ½" diameter in boxes. There's nothing complicated about using an electric drill, but if you're new to this tool, have someone at your local home improvement center walk you through it.

Here's an easy way to decide what size drill bit to use. Try slipping drill bits through the nut used to secure the screws or electrical component you're drilling the hole for. Choose a drill bit that's too big to fit through the hole in the nut and smaller than the outside of the nut.

Drill bits sometimes *bind* in the material you're drilling. When a drill bit binds, the box gets kind of edgy and begins to spin with the drill. That's why it's important that you clamp the box you're drilling to your worktable or secure it in a vise. We've found that drill bits bind more often in plastic boxes than in wooden ones.

Mounting your project in a box

After you build your circuit and drill or cut holes in your box to accommodate anything you want to feed through from inside to outside, actually mounting things in the box has a few ins and outs, too.

Working with switches, potentiometers, and other panel-mount components

Many switches, potentiometers, and other components have a threaded shaft, a nut, and possibly washers that are meant to be mounted through a hole in a panel. Here's the drill (pun intended):

1. **Drill a hole in your box where you want to mount the component.**

2. **Clean up any debris around the hole from the drilling.**

3. **Slide the threaded shaft through the hole and tighten a nut on the threads.**

   If the hole turns out to be a little too big, slip a washer under the nut.

Some components, such as speakers or buzzers, have holes in flanges that you can use to secure the component to the wall of the box with screws. Use the flange holes as a template to mark the locations to drill holes.

Some components are meant to be panel-mounted but don't have threads. For example, in Chapter 11, we use a two-piece, LED panel-mount socket in which one part of the socket slides through the hole and onto the other half and snaps in place, securing the LED.

In some cases, you have to mount microphone cartridges that don't come with threads or snap sockets in the walls of boxes. Simply drill a hole that's just big enough in diameter to slip the cartridge in with a snug fit. If you're using a wooden box, the wall of the box should be thick enough to secure the microphone cartridge. If you're using a plastic box, you'll probably need to secure the microphone cartridge with glue.

Sticking things on the box

If you use screws to attach a component to a wooden box and you want to put something else over the screw head, use a flathead screw, as shown in Figure 4-16. When you don't need to keep the surface flush, panhead screws are just fine.

Figure 4-16:
Keep the surface flush with flathead screws.

When stranded wire works

If you're connecting a component in the lid of your box to a terminal block on the breadboard in the bottom of the box, that wire will be bent back into the box when you close the lid. In this case, it's best to use stranded wire, which is more flexible than solid wire. (Refer to Figure 4-9 for a comparison.)

Be sure to leave enough length of wire for the box to open and lightly solder the strands at the end of the wire together before inserting them into the terminal block so none of them poke out and short the circuit.

For items with flat surfaces such as breadboards or battery packs, Velcro or similar materials are useful to secure them in your box. However, if you remove breadboards that you have secured in a box with Velcro, be careful. Breadboards have a thin sheet of plastic or paper on the back that can peel off if you're not careful.

You can make custom mounts out of metal bars, screws, and nuts, as shown in Figure 4-17, which shows also how we mount the motors in Chapter 13.

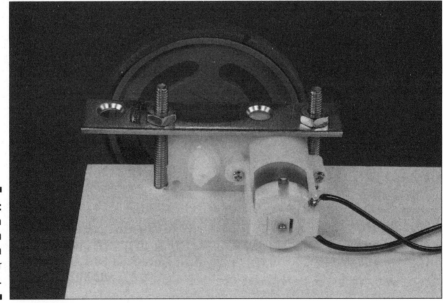

Figure 4-17:
Make a custom mount with a metal bar and screws.

You can also use wooden dowels and cable ties as we did to mount the microphone in Chapter 6 (see Figures 4-18 and 4-19).

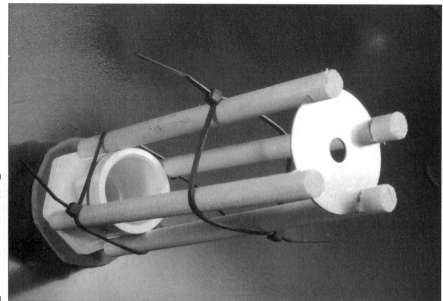

Figure 4-18:
Make a custom mount with wooden dowels and cable ties.

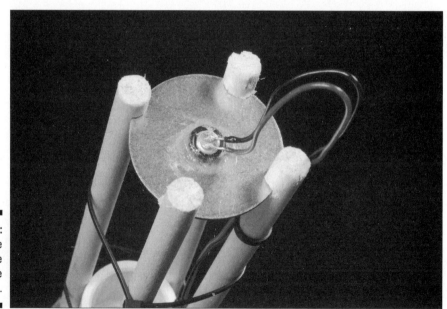

Figure 4-19:
With the microphone cartridge in place.

You can use wire clips to secure wires to the side of boxes so they are out of the way as you work, as shown in Figure 4-20.

Figure 4-20:
Use wire
clips to
secure the
wires.

Part II
Sounding Off!

In this part . . .

*W*e thought we'd start you working on projects that provide a bang, so this first set of projects is all about making noise. The chapters in this part set lights dancing to music and help you pick up sounds at a distance with a parabolic microphone. You also get to build your own AM radio.

Along the way, you'll also pick some information about working with sound synthesizer chips as well as using components to detect various audio frequencies, microphone sensitivity, and more.

Chapter 5

Making Light Dance to the Music

Music hath charms to soothe the savage breast, as the saying goes. Whether you feel particularly savage or not, if you enjoy listening to all kinds of music, you'll enjoy this project. Here, we combine light and sound in an interesting way: We show you how to set up two rows of lights that illuminate to different frequencies of sound. When you put on a rowdy piece of music — say, swing or reggae — the lights dance all around. Every piece of music has its own, unique effect.

We chose to arrange our lights like notes on a music staff, but you could put them in any arrangement you like. By working through this fun project, you get to know more about frequency filters, operational amplifiers, and how music can make things light up and get your toes tapping.

The Big Picture: Project Overview

After you complete this project, you'll have a display of LEDs that light up in response to high- or low-frequency sounds. You can see our LED musical notes arrangement in the finished display in Figure 5-1.

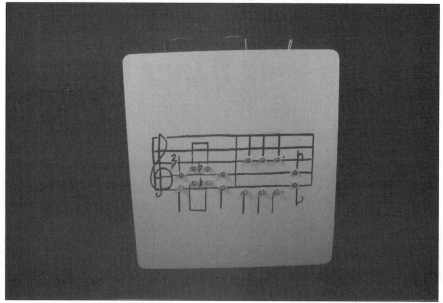

Figure 5-1:
The final
product.
Dance to
the music!

Here's the overview of what you'll be up to in the Dance to the Music project. You will

- **Put together an electronic circuit to turn on the LEDs in response to sounds.**

    Half of the LEDs will flash to high-frequency sounds, and the other half will flash to low-frequency sounds.

- **Create a template for the musical notes; then place the template on the box and drill holes in the top of a wooden box for the LEDs.**

- **Wire two groups of LEDs and resistors.**

- **Turn on the juice (that is, pop in the batteries and flip the switch), and then turn on some music.**

    The circuit sends current to each group of LEDs in response to the music.

- **Get dancing yourself.**

    This one's infectious!

Scoping Out the Schematic

Music and light don't just happen: You have to start with a plan — or in this case, a schematic. You get to put together one large breadboard and two LED arrays for this project.

Take a look at the schematic for the breadboard shown in Figure 5-2.

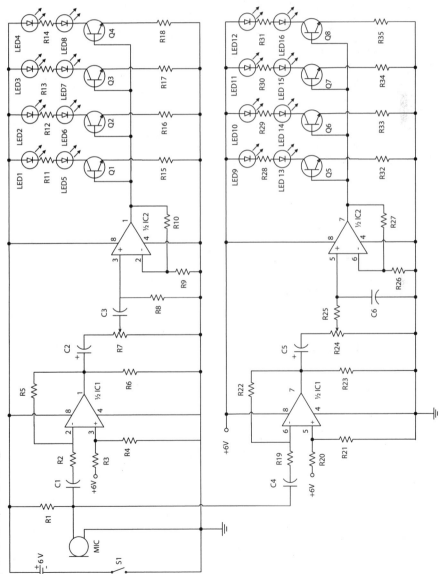

Figure 5-2:
The schematic for your Dance to the Music project.

The following sections explore the schematic elements in detail.

Fancy Footwork: Exploring the Dance to the Music Circuit

To make your musical score light up in response to the music, you need to make a circuit that uses a microphone, as well as two operational amplifier ICs in combination with some resistors, capacitors, and transistors. Working together, these control which group of LEDs lights up to a particular frequency of music.

Here's the overview of the schematic elements that you use to control your project:

- ✔ The circuit starts with an electret microphone, which transforms sound into electrical signals.
- ✔ **R1** connects the microphone to positive voltage and supplies about 4.5 volts required by the microphone to function.
- ✔ **C1 and C2** are capacitors that block the DC voltage on the input signal and allow the AC signal to pass.
- ✔ At this point, the circuit splits: The signal processed by the upper half of the circuit powers the LEDs blinking in response to high-frequency sound; the signal flowing through the lower half of the circuit powers the LEDs blinking in response to low-frequency sound.
- ✔ **IC1** is an operational amplifier (op amp) that amplifies the signal from the microphone. The IC contains two op amps; one half of IC1 is used in the upper circuit, and one half of IC1 is used in the lower circuit.
- ✔ **R2 and R5** in the upper circuit and **R19** and **R22** in the lower circuit set the gain for each side of IC1. Because R5 is 50 times R2, a signal processed by the op amp is amplified approximately 50 times.

  Gain is the amplitude of the voltage out divided by the amplitude of the voltage in: in other words, how much more juice goes out than comes in.

- ✔ **R3, R4, and R6** in the upper circuit and **R20, R21,** and **R23** in the lower circuit provide a DC bias to the op amp that allows the full AC signal to be amplified. If these resistors were not there, the portion of the AC signal coming into the op amp with voltage less than 0 volts would be lost.

Filters

The portion of the circuit made up of C3 and R8 is an *RC high pass filter*. This pass has nothing to do with a tricky move in football. Rather, with this filter, signals above a certain frequency — determined by the value of C3 and R8 — pass through more easily than signals below that frequency. The strength of signals below this key frequency is therefore reduced. This type of filter is made up of the capacitor in line with the signal path and the resistor between the output of the signal and ground.

Correspondingly, the portion of the circuit made up of R25 and C6 is an *RC low pass filter*. Here,

signals below a certain frequency — determined by the value of R25 and C6 — pass through more easily than signals above that frequency. The strength of signals above this key frequency are reduced. This type of filter has the resistor in line with the signal path and the capacitor between the output of the resistor and ground.

For both types of filters, increasing the value of either or both the resistor or capacitor lowers the value of the key frequency. Lowering the value of either or both the resistor or capacitor increases the value of the key frequency.

Bias involves applying voltage that is above ground to a portion of the circuit to amplify both the positive and negative sides of a signal. Without DC bias, you would lose part of the signal.

- **C2 and C5** remove any DC bias from the signal coming out of the op amps.

- **R7 and R24** are potentiometers that allow you to adjust the sensitivity of the circuit in relation to how loud the music is.

- **C3 and R8** function as a high pass filter, and **R25 and C6** function as a low pass filter. These filters are what make higher-frequency sounds light up LEDs 1–8 and lower-frequency sounds light up LEDs 9–16.

- **IC2** is an op amp that is used to amplify the signal that passes through the filter. The IC contains two op amps; one half of IC1 is used in the upper circuit, and one half of IC1 is used in the lower circuit.

- **R9 and R10** in the upper circuit and **R26 and R27** in the lower circuit set the gain of the op amp. Because R10 is 200 times R9, a signal processed by the op amp is amplified approximately 200 times.

- **Q1, Q2, Q3, and Q4** in the upper circuit and **Q5, Q6, Q7, and Q8** in the lower circuit are 2N3904 transistors whose bases are connected to the output of the op amps in IC2. When the output of the op amp reaches about 0.7 volts, the transistors turn on, and current flows through the LEDs.

The circuit won't mean a thing if you don't set up the lights for it to control. That's where the two groups of LEDs and resistors come in. They include

- **LED1–LED8** to light the display for the high-frequency circuit and **LED9–LED18** to light the display for the low-frequency circuit.

- **R11–R18** in the upper circuit and **R28–R35** in the lower circuit are resistors, that in series with LEDs, limit the current running through the LEDs to approximately 10 milliamps.

Building Alert: Construction Issues

Behind the musical score, we drew on the box that holds the circuit, and all the resistors and LED leads are soldered together. To make sure that leads that get bent and touch don't cause a short, they need to be protected. Rather than using electrical tape, we use liquid electrical tape to coat the exposed leads.

Wires run between the circuit board resting on the bottom of the box and the LEDs that you insert in the top of the box. Be sure to cut your wires long enough so that when you open the box (which moves the LEDs in the top farther away from the circuit in the bottom of the box) that you don't rip out the wires. Also be careful that you leave room to tuck those long wires inside the box when closed so they don't poke out the sides or get caught in the hinges. This is one case when stranded wires work better than solid wires because they are more flexible.

Perusing the Parts List

It's off to your nearest electronics store or online vendor for those electronic parts you use to build the circuit and assemble the box that contains all those LEDs.

The circuit that transforms music into your dancing light show involves the following parts, several of which are shown in Figure 5-3:

✔ **2.2 kohm resistor (R1)**

✔ **Eight 220 ohm resistors (R11–R14, R28–R31)**

✔ Eight 100 ohm resistors (R15–R18, R32–R35)

✔ Two 10 kohm potentiometers (R7, R24)

✔ Four 47 kohm resistors (R3, R4, R20, R21)

✔ Two 100 kohm resistors (R5, R22)

✔ Two 2 kohm resistors (R2, R19)

✔ Three 5 kohm resistors (R6, R8, R23)

✔ Two 1 kohm resistors (R9, R26)

✔ One 10 kohm resistor (R25)

✔ Two 220 kohm resistors (R10, R27)

✔ 0.001 microfarad ceramic capacitor (C3)

✔ Three 0.1 microfarad ceramic capacitors (C1, C4, C6)

✔ Two 10 microfarad electrolytic capacitors (C2, C5)

✔ Eight green size T-1 ¾ LEDs (LED1–LED8)

✔ Eight red or orange size T-1 ¾ LEDs (LED9–LED16)

✔ Two LM358 op amps (IC1 and IC2)

✔ Electret microphone

We use the Horn part #EM9745-38 electret microphone for two reasons: high sensitivity and a reasonable size that makes it easy to handle. You can use other electret microphones; see Chapter 3 for the criteria to help you choose one.

✔ Two 830-contact breadboards

✔ One 4 AA battery pack with snap connector

✔ Eleven 2-pin terminal blocks

✔ Two knobs (for the potentiometer)

✔ Eight 2N3904 transistors (Q1–Q8)

✔ A wooden box

We found one at a local craft supply store that was just the right size to hold the electronics for this project.

✔ An assortment of different lengths of prestripped, short 22 AWG wire

✔ Several feet of black 20 AWG wire

✔ Several feet of red 20 AWG wire

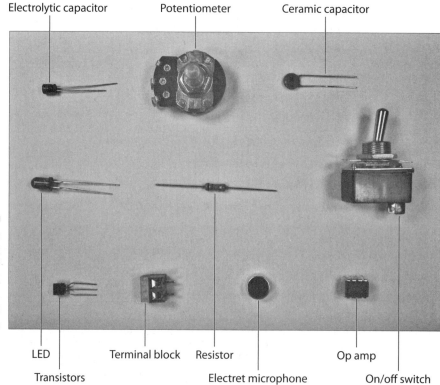

Electrolytic capacitor   Potentiometer   Ceramic capacitor

Figure 5-3:
Important
pieces of
the Dance
to the Music
project

LED   Terminal block   Resistor   Op amp

Transistors   Electret microphone   On/off switch

Taking Things Step by Step

To make the lights dance to the music, you have a few tasks to perform:

1. Build the circuit that makes it all run.

2. Assemble the lights on the surface of the wooden box that you place the circuit in.

3. Attach knobs, an on/off switch, a microphone, potentiometers, and a speaker to the box; then insert the breadboard containing the circuit in the box and connect everything.

And that's what we cover in this section.

Building a circuit

Time to go one-on-one with your breadboard. Grab your schematic, tape it to the wall, and get going.

Here are the steps involved in building the circuit:

1. **Place the two LM385 ICs (IC1 and IC2) and 11 terminal blocks on the breadboard, as shown in Figure 5-4.**

   As you can see in this figure, each terminal block has connections for two wires. Here's what the wires get from each block will be connected to:

   - Wires from one of the terminal blocks on the right side of the figure go to the battery pack.

   - Wires from the other terminal block on the right go to the microphone.

   - Wires from the terminal blocks in the center go to the potentiometers.

   - Wires from the terminal blocks on the left side of the breadboard go to the LEDs.

2. **Place the eight 2N3904 (Q1–Q8) transistors on the breadboard, as shown in Figure 5-4.**

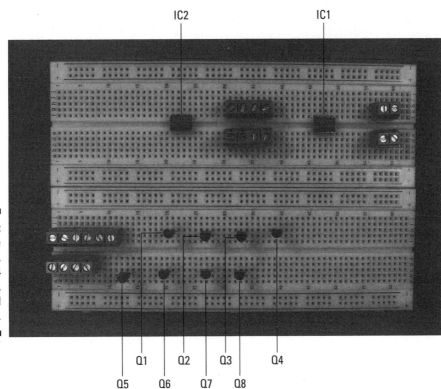

Figure 5-4:
Place the LM385 ICs, 2N3904 transistors, and terminal blocks.

Insert each transistor lead into a separate breadboard row with the collector lead to the left side (as shown in Figure 5-4), the base lead in the center, and the emitter lead to the right. The pin designations of the 2N3904 transistors are shown in Figure 5-5.

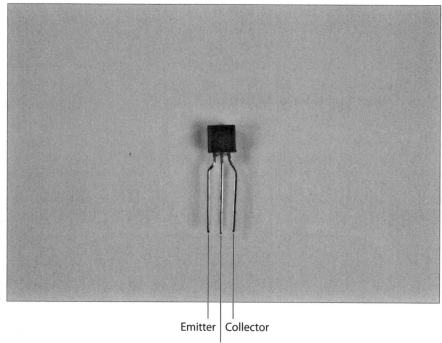

Figure 5-5:
The 2N3904
transistor
pinout.

Emitter | Collector

Base

3. **Insert wires to connect the ICs, the battery pack terminal block, the microphone terminal block, and the potentiometer terminal blocks to the ground bus. Then insert wires between the ground buses to connect them to each other, as shown in Figure 5-6.**

The ground buses are designated by a negative (–) sign on this breadboard.

4. **Insert wires to connect the ICs, the battery terminal block, and one of the LED terminal blocks to the +V bus. Then insert wires between the +V buses to connect them, as shown in Figure 5-7.**

The +V buses are designated by a + sign on this breadboard.

5. **Insert wires to connect the ICs, terminal blocks, and discrete components, as shown in Figure 5-8.**

Figure 5-6:
Shorter wires connect components to ground bus; the two long wires on the right connect the ground buses.

LED terminal block to +V

Pin 8 of IC2 to +V

Battery terminal block to +V

Pin 8 of IC1 to +V

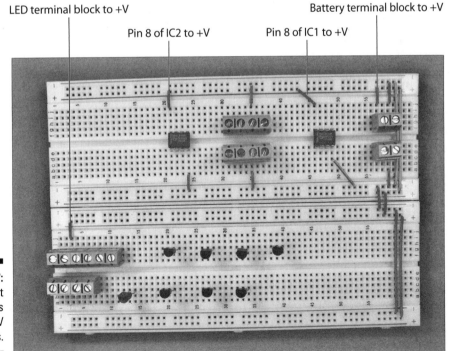

Figure 5-7:
Connect components to the +V bus.

Potentiometer R24 terminal block to open region   Pin 7 of IC1 to open region

Pin 5 of IC2 to open region   Pin 6 of IC1 to open region

Pin 6 of IC2 to open region   Connecting two rows

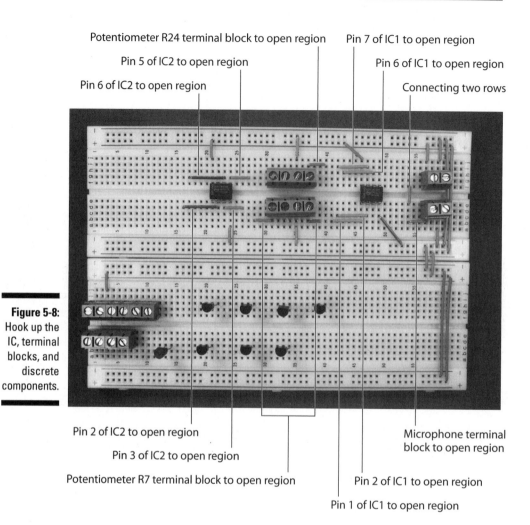

Figure 5-8:
Hook up the
IC, terminal
blocks, and
discrete
components.

Pin 2 of IC2 to open region   Microphone terminal
block to open region

Pin 3 of IC2 to open region

Potentiometer R7 terminal block to open region   Pin 2 of IC1 to open region

Pin 1 of IC1 to open region

6. **Insert wires to connect IC2 to the transistors and the base pins of the transistors to each other, as shown in Figure 5-9.**

7. **Insert three 0.1 microfarad capacitors (C1, C4, and C6), two 10 microfarad capacitors (C2 and C5), and one 0.001 microfarad capacitor (C3) on the breadboard, as shown in Figure 5-10.**

   Use both the schematic and the photo to place each component. For example, the schematic shows that the + side of C2 is connected to Pin 1 of IC1 and that the other side of C2 is connected to potentiometer R7, so insert the long lead of C2 in the same row as the wire connected to pin 1 of IC1 and the short lead in the same row as the wire connected to the terminal block for potentiometer R7.

Pin 7 of IC2 to base of Q5

Pin 1 of IC2 to base of Q1

Base of Q1 to base of Q2

Base of Q2 to base of Q3

Base of Q3 to base of Q4

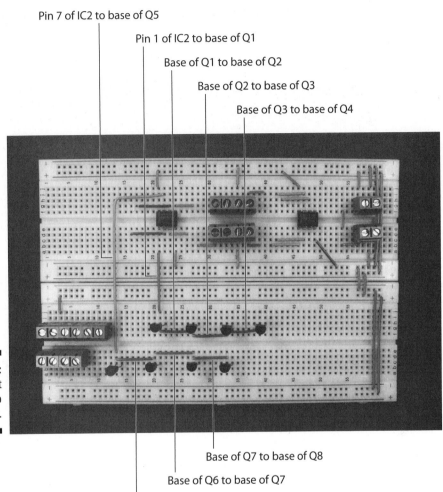

Figure 5-9:
Connect
IC2 to
transistors.

Base of Q7 to base of Q8

Base of Q6 to base of Q7

Base of Q5 to base of Q6

8. **Insert eight 100 ohm resistors (R15–R18, R32–R35) from the emitter pins of each transistor to the ground bus, as shown in Figure 5-11.**

9. **Insert one 2.2 kohm resistor (R1), one 2 kohm resistor (R2), two 47 kohm resistors (R3 and R4), one 100 kohm resistor (R5), two 5 kohm resistors (R6 and R8), one 1 kohm resistor (R9), one 5 kohm resistor (R6), and one 220 kohm resistor (R10) on the breadboard, as shown in Figure 5-12.**

C4 from microphone TB to open region

C5 from IC1 Pin 7 to R24 TB

C6 from IC2 Pin 5 to ground bus

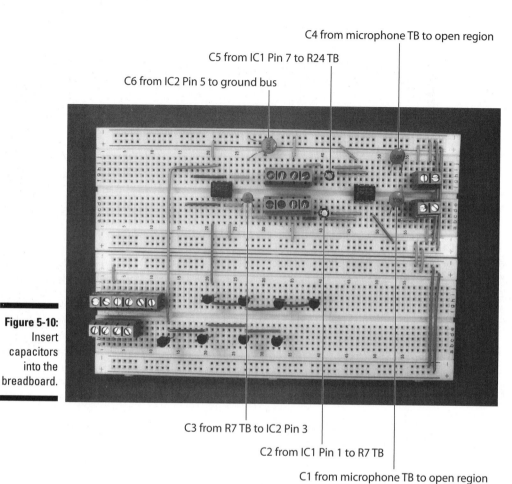

Figure 5-10:
Insert
capacitors
into the
breadboard.

C3 from R7 TB to IC2 Pin 3

C2 from IC1 Pin 1 to R7 TB

C1 from microphone TB to open region

10. **Insert one 2 kohm resistor (R19), two 47 kohm resistors (R20 and R21), one 100 kohm resistor (R22), one 5 kohm resistor (R23), one 10 kohm resistor (R25), one 1 kohm resistor (R26), and one 220 kohm resistor (R27) on the breadboard, as shown in Figure 5-13.**

11. **Insert wires to connect the collector pins of the transistors to the terminal blocks, as shown in Figure 5-14.**

R15   R16   R17   R18

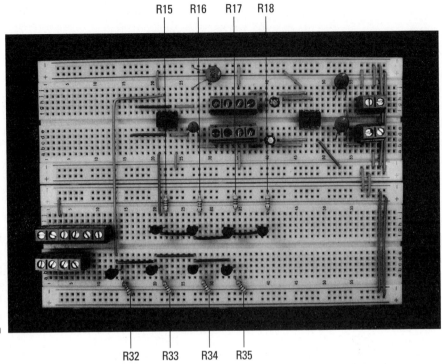

Figure 5-11:
Insert
resistors
onto the
breadboard.

R32   R33   R34   R35

Let there be lights

All the brains of the circuit assembled in the previous section are there to make the Dance to the Music light display work. Here's the part where you assemble those lights.

Follow these steps to create your Dance to the Music display:

1. **Select a series of musical notes, with eight high notes and eight low notes.**

2. **Draw a musical staff on the top of the wooden box in pencil and then draw a dot for the spot where each LED will go.**

3. **Drill test holes in a piece of scrap wood to determine the size of drill bit that you should use to give a press fit for the LEDs.**

   We used a ¹³⁄₆₄" drill bit.

R1 from C1 to +V bus

R2 from C1 to Pin 2 of IC1

R5 from Pin 1 of IC1 to Pin 2 of IC1

R10 from Pin 1 of IC2 to Pin 2 of IC2

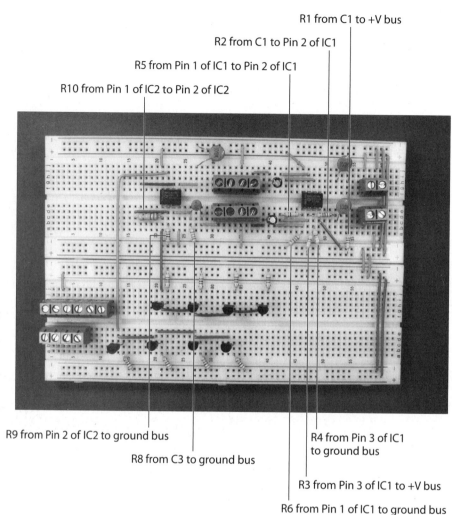

Figure 5-12:
Insert more
resistors
onto the
breadboard.

R9 from Pin 2 of IC2 to ground bus

R8 from C3 to ground bus

R4 from Pin 3 of IC1
to ground bus

R3 from Pin 3 of IC1 to +V bus

R6 from Pin 1 of IC1 to ground bus

4. **Drill holes for the LEDs at the locations you marked in Step 3.**

5. **With a permanent marker or paint brush, draw the musical notes and staff on the wooden box.**

6. **Insert LEDs in the drilled holes.**

   The box at this stage is shown in Figure 5-15.

   See Figure 5-15 for help with this if you're not musical!

R21 from Pin 5 of IC1 to ground bus

R20 from Pin 5 of IC1 to +V bus

R23 from Pin 7 of IC1 to ground bus

R26 from Pin 6 of IC2 to ground bus

R27 from Pin 7 of IC2 to Pin 2 of IC2

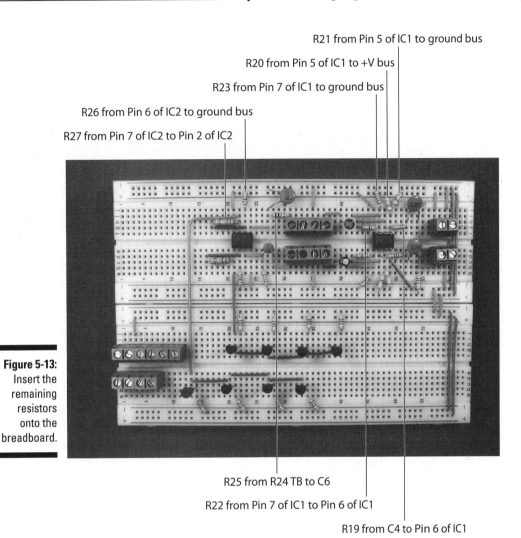

Figure 5-13:
Insert the
remaining
resistors
onto the
breadboard.

R25 from R24 TB to C6

R22 from Pin 7 of IC1 to Pin 6 of IC1

R19 from C4 to Pin 6 of IC1

7. Attach resistors between the pairs of LEDs, as shown in Figure 5-16.

Attach the resistors to the short lead on one LED of each pair (four pair of each color) and to the long lead on the other LED of the pair.

8. Solder the resistors to the leads and clip the leads just above the solder joint.

Clip only the leads to which you have soldered resistors. Figure 5-16 shows how the LEDs and resistors should look at this point. Figure 5-17 shows a close up to help you see the soldering more clearly.

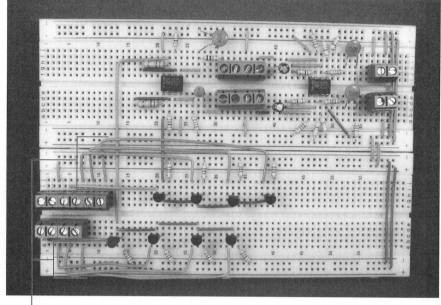

Figure 5-14:
Connect the collector pins of the transistor to terminal blocks.

Wires connecting transistor collector pins to terminal blocks

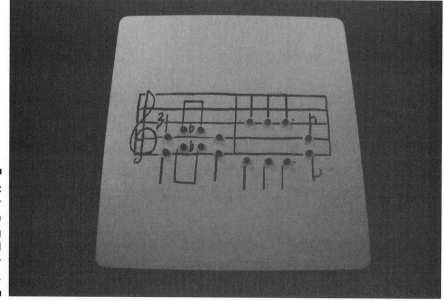

Figure 5-15:
A bit of our favorite song created from marker and LEDs.

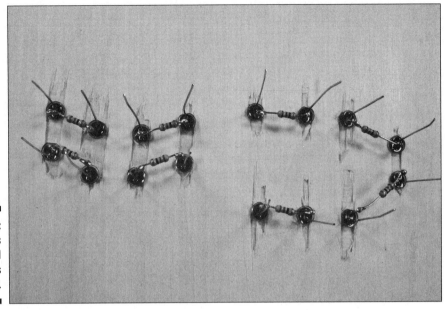

Figure 5-16:
Resistors
soldered
and leads
clipped.

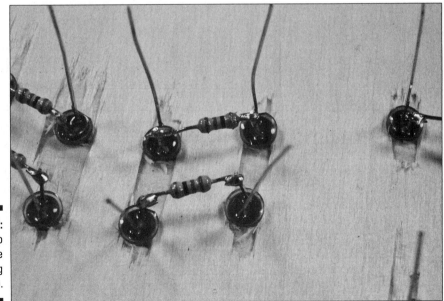

Figure 5-17:
A close-up
of the
soldering
job.

Be sure to heed all the safety precautions about soldering that we provide in Chapter 2. For example, don't leave your soldering iron on if you have to step away to let the pizza delivery guy in. And just in case a bit of solder has an air pocket that could cause it to pop, wear your safety glasses whenever you solder or clip leads and wires.

9. **Connect and solder short lengths of 20 gauge red wire between the remaining long LED leads. Then attach a 12" 20 gauge red wire to the remaining long lead of the first LED pair, as shown in Figure 5-18.**

   These wires are the +V bus for the LED arrays; you attach the 12" wire to a terminal block to supply voltage to the bus.

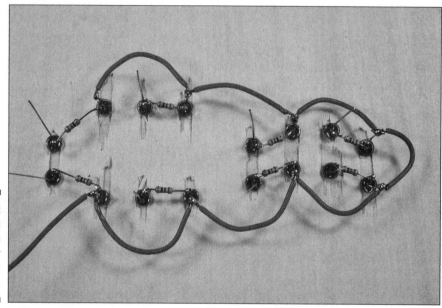

Figure 5-18: Red wires forming the +V bus for the LED arrays.

10. **Clip the LED leads just above the solder joint.**

    Clip only the leads to which you have soldered the red wires; you need the remaining leads in the next step.

11. **Connect and solder a 12" 20 gauge black wire to the remaining short LED lead on each pair of LEDs, as shown in Figure 5-19.**

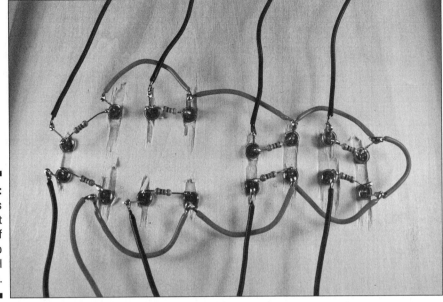

Figure 5-19: Black wires to connect each pair of LEDs to terminal blocks.

You will connect each of these eight wires to a terminal block. The upcoming Figure 5-20 shows these connections close up.

Figure 5-20: A close-up of all the connections.

12. **Clip the LED leads just above the solder joint.**

Make sure that the LED leads and solder joints don't touch each other. Coat them with liquid electrical tape to help prevent any shorts that could occur if you bend or push the wires together.

Adding the rest of the doohickeys

After the circuit and LEDs are pretty much taken care of, you have a lot of other things still sitting on your workbench, like the microphone, potentiometers, a switch, and one or two more things. You have to take care of assembling these before your project will work.

Follow these steps to take care of all the rest of the parts that make the music dance:

1. **Solder a 12" black wire to the microphone ground contact and a 12" red wire to the microphone +V contact.**

   Figure 5-21 identifies the microphone contacts.

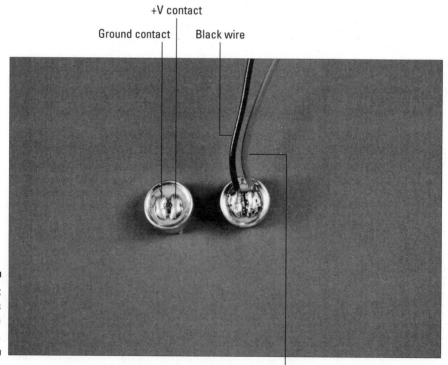

+V contact

Ground contact   Black wire

Figure 5-21:
Solder wires
to the
microphone.

Red wire

2. **Drill holes in the box where you will insert the microphone and on/off switch.**

   We put both the potentiometers and on/off switch on one side of the box and the microphone on another side, but the placement is really up to you. Chose a drill bit size for the microphone hole so that the microphone has a slip fit. The upcoming Figure 5-22 shows where we place these components.

   See Chapter 4 for more information about choosing drill bit sizes for particular components and other pieces of wisdom on how to customize a box for your projects. Make sure you use safety glasses when drilling, and clamp the box to your worktable!

3. **Slip the shaft of the on/off switch through the drilled hole and secure with the nut provided.**

4. **Slip the shaft of the potentiometers through the drilled holes and secure with the nuts provided.**

5. **Slip a knob over the shaft of each potentiometer and tighten the knob with the set screw provided.**

   The tread on potentiometers is about ¼" long, so if the wall of your wooden box is ¼" thick, you won't be able to use the nut to secure the potentiometer. Instead, check to make sure that the potentiometer shaft extends far enough beyond the box to allow the knob set screw to tighten on the shaft. If the shaft extends far enough, glue the face of the potentiometer to the box, making sure that you don't get any glue on the rotating shaft of the potentiometer. If the shaft doesn't protrude quite enough, use a small chisel to remove some wood on the inside of the box to let the potentiometer shaft extend a little farther before you glue.

6. **Slip the microphone into its drilled hole with a press fit.**

   Figure 5-22 shows the on/off switch, potentiometers, and microphone in place in the box.

7. **Solder 12" black wires to each of the three potentiometer lugs, as shown Figure 5-23.**

8. **Solder the black wire from the battery pack to one lug of the on/off switch and solder a 12" black wire to the remaining lug of the on/off switch.**

   Figure 5-23 shows the switch after the wires are soldered.

9. **Attach Velcro to the breadboard and the box and then secure the breadboard in the box.**

10. **Attach Velcro to the battery pack and the box and then secure the battery pack in the box.**

On/off switch   Microphone

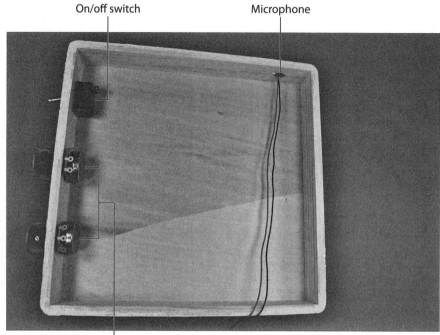

Figure 5-22:
Box with on/
off switch,
potentio-
meters, and
microphone
in place.

Potentiometers

11. **Insert the wires from the LEDs, battery pack, and the on/off switch to the terminal blocks on the breadboard, as shown in Figure 5-24.**

Use the following as a key to the numbered callouts in Figure 5-24.

      1. Red wire from LED +V bus

    2–5. Wires from pair of red LEDs

    6–9. Wires from pair of green LEDs

    10. Red wire from battery pack

    11. Black wire from on/off switch

    12. Red wire from microphone

    13. Black wire from microphone

14 and 17. Wires from right potentiometer lug

15 and 18. Wires from center potentiometer lug

16 and 19. Wires from left potentiometer lug

Black wire from battery pack to on/off switch

Black wire to on/off switch

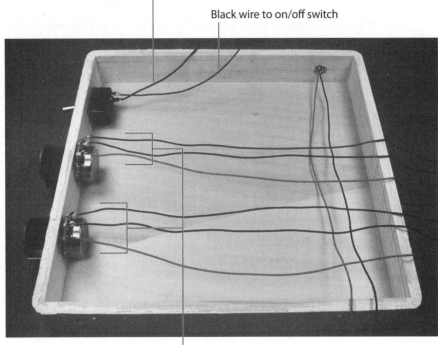

Figure 5-23:
Wires
soldered to
the on/off
switch and
the potenti-
ometers.

Wires to potentiometers

12. **As you insert the wires, cut each of them to the length needed to reach the assigned terminal block and strip the insulation from the end of the wire.**

13. **Secure the wires with wire clips.**

Trying It Out

Okay, you've been taking it on faith that this is going to be a cool project, and we appreciate it. But now it's time to get the thing going and see whether you think it's as much fun as we do.

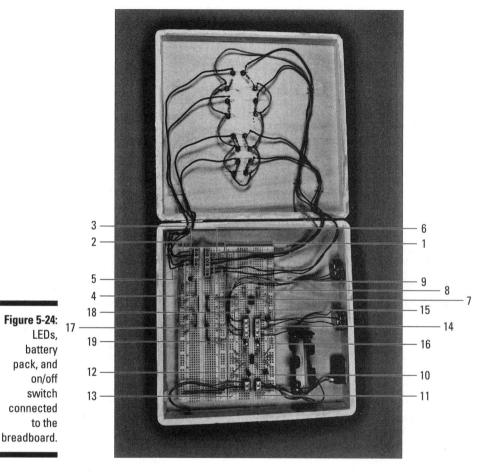

Figure 5-24: LEDs, battery pack, and on/off switch connected to the breadboard.

You can use any music you like to get the effect going, but we found that music with lots of instruments — like a swing band with lots of brass — works best. Also, music with an upbeat tempo moves along and gets the lights switching on and off faster, which makes the effect better. Our favorite number for dance to the music? Ella Fitzgerald singing *Take the A Train,* by Billy Strayhorn. (Ask your parents; they might have heard of it.)

Here are the simple steps to get this project going:

1. **Pop the batteries into the battery pack.**

2. **Flip the on/off switch to on.**

3. Put on some music.

That's it! Watch the lights go on an off in response to the high and low frequencies in the music. You can adjust the sensitivity of the LEDs by turning the potentiometers.

Here are the obvious things to check out if you're having a problem:

✔ Check that all the batteries are fresh and tight in the battery pack and that all face the right direction.

✔ If one or two LEDs aren't working, replace them.

✔ If two LEDs in series with each other aren't functioning, you might have reversed the long and short leads of the LEDs; if so, just replace that pair of LEDs.

✔ You are playing *Brahm's Lullaby*. *Brahm's Lullaby* will not light up a single LED. Switch to Snoop Doggy Dogg or Motörhead.

Taking It Further

By now, you're probably jumping and jiving to this cool light show, playing every CD you have to see what they do, and still, you want more? Here are a few different ways to take this project further:

✔ Obviously, you can change from a musical staff and notes to any kind of shape you might want to define with your LEDs. You can have two stars or a sun and moon, for example.

✔ You can use a *band pass filter* to add more layers of frequency. For example, between your high pass filter and low pass filter, you could add two more band pass filters to hit intermediate frequencies and have four sets of LEDs going off in response to music.

✔ You could miniaturize your circuit so you can pin it on your shirt or take it with you to parties. A few steps would help you to get a smaller circuit. First, you could use smaller LEDs. (This project uses T-1 ¾ LEDs, but you could use T1 LEDs.) You could also use a different method of building the circuit called a *dead bug circuit*. Imagine an IC turned on its back with its little prongs sticking up on the air, and you get an idea of what we mean. This method doesn't involve a breadboard but makes connection directly to the LED. Check out www.arrl.org (American Radio Relay League online) for some ideas about using the dead bug approach to building circuits.

Chapter 6

Focusing Sound with a Parabolic Microphone

*W*hen's the last time you stepped outside your door and really listened to all that noise out there? There are birds, traffic noises, maybe a cat slinking through the tall grass. Of course, some of these sounds you can pick up with your own two ears, but others aren't so clear. A microphone that could help you pick up those softer sounds — whether made by your buddy whispering to you 100 feet away or birds in the trees — might come in handy.

In this project, we show you how to put together a microphone and an IC to amplify noises that it picks up. Then you can put the whole circuit together with a *parabolic* (curved) metal dish that picks up sounds, just like how cupping your hand behind your ear helps you pick up sounds.

What a Dish! The Project Overview

A parabolic microphone kind of looks like a TV satellite dish. Instead of picking up electromagnetic broadcast waves, though, this kind of microphone picks up sound waves. *Sound waves* are simply vibrating molecules that deliver all kinds of sounds, like birds chirping and your neighbors' latest argument. (*Disclaimer:* Of course, we're NOT recommending that you use this project to eavesdrop on anybody!) You use the microphone, the curved dish, and headphones to gather and deliver the amplified sound to your ears.

You can see the finished parabolic microphone (and our own author, Earl, lurking behind it) in Figure 6-1.

Figure 6-1:
Parabolic
microphone
and ace
operator.

Here are the types of activities that you'll be doing to create your own parabolic microphone. You will

1. Put together

    • An electronic circuit containing an electret microphone cartridge that can pick up faint sounds

    • An IC that amplifies the sounds enough to power your headphones

2. Install the microphone in a parabolic dish that gathers sound like a giant ear.

We use a parabolic dish because the shape gathers sound and focuses it on a point — in this case, on the spot where we put the microphone cartridge. When you point the parabolic dish at a bird a hundred feet away, for example, the dish gathers sound only from the exact direction you point toward and focuses the sound on the microphone cartridge.

3. Add a handle made of pieces of plastic pipe so you can hold the dish up easily and aim it at the feathered or non-feathered friend of your choosing.

Scoping Out the Schematic

By now, your mind is likely a-flutter with ideas for sounds floating around your neighborhood that you're dying to pick up, so it's time to get going on the schematic for the board, which you can see in Figure 6-2.

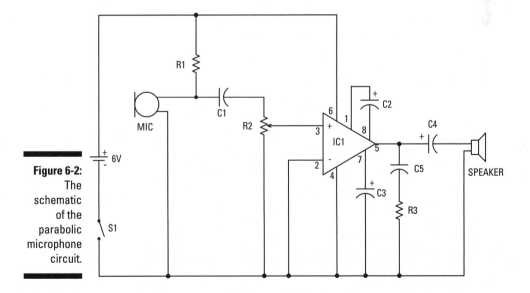

Figure 6-2: The schematic of the parabolic microphone circuit.

The following is a list of the schematic elements for your parabolic microphone:

- The circuit starts with the electret microphone, which transforms sound waves into electrical signals.

- **R1** connects the microphone to positive voltage and supplies the 3 volts required to make the microphone function.

- **C1** is a capacitor that blocks the DC voltage on the input signal and allows the AC signal to pass.

- **IC1,** an LM386N-1 audio amplifier, takes the electrical signal generated by the electret microphone and amplifies it to provide sufficient power to generate sound through the headphones.

- **R2** is a potentiometer you use to control the sound volume.

- **C2** sets the voltage gain of IC1 to 200. Therefore the voltage out is 200 times the voltage in.

- **C3** improves the stability of the LM386 amplifier to prevent problems, such as oscillation. Oscillation can turn that nice little birdcall into a hodgepodge of unintelligible sounds.

- **C4** removes any DC offset from the output of the LM386 amplifier.

- **C5** acts as a current bank for the output. This capacitor drains when sudden surges of current occur and refills with electrons when the demand for current is low.

Building Alert: Construction Issues

The electret microphone cartridge used in this project is about ¼" in diameter. Therefore, the contact pads on the cartridge to which you solder wires are very small. Review the methods for soldering to contact pads that we cover in Chapter 4. And don't forget to use your magnifying glass after soldering to check for *solder bridging* (a clump of solder that could short out the connection) between the two pads.

If you find bridging solder, try using a utility knife to ***gently*** clean it out. Also, you might subscribe to the better-safe-than-sorry theory and consider ordering two of these microphone cartridges. That way, if you ruin one while learning how to solder to these small contact pads, you have a second one at the ready, saving you time and shipping fees for a second order.

The parabolic dish comes with a reflective finish. We recommend spray painting it to avoid blinding yourself or others with the sun's reflection from the dish's original finish. ***Note:*** Before painting, be sure to wipe off the dish to get rid of dust or smudges because spray painting makes any smudges on the smooth surface really stand out.

Feel free to get creative when choosing a color for the dish; if you're planning to listen to wildlife, you might want to paint the dish a color that will blend in, such as green or brown. We painted ours black because, well, we had the black paint.

Use a special glue — PVC cement to connect the PVC fittings that form the microphone handle. You can get this type of glue at any building supply store. This glue provides a very strong joint, which you need because the dish is somewhat heavy and you don't want the handle to fall apart, causing you to drop the dish on the ground. (This happened to us with a prototype we hadn't yet glued together. Nancy's ears are still ringing from the resulting amplified crash.)

Be sure to wear some form of work gloves when using this glue because it melts plastic — you definitely don't want this stuff on your hands! Also read the label for advice on safety, such as using the glue in a well-ventilated area as well as what to do if it does come into contact with your skin.

Perusing the Parts List

Here comes our favorite part — shopping for parts! The upcoming Figures 6-3 and 6-4 show many of the parts used in this project.

Here's what you'll need:

- ✔ **Electret microphone cartridge**

  Trust us. We tried a lot of microphones to get this working just right (and save you the trouble). We finally decided on the Panasonic part #WM61A because of its very high sensitivity that allows it to pick up faint sounds. We found the WM61 at Digikey (www.digikey.com).
- ✔ **10 kohm potentiometer (R2)**
- ✔ **10 ohm resistor (R3)**
- ✔ **5.6 kohm resistor (R1)**
- ✔ **0.1 microfarad ceramic capacitor (C1)**
- ✔ **0.047 microfarad ceramic capacitor (C5)**
- ✔ **10 microfarad electrolytic capacitors (C2, C3)**

✔ **100 microfarad electrolytic capacitor (C4)**

✔ **LM386N-1 amplifier IC1**

Of the many versions of the LM386 amplifier, we chose the LM386N-1 because it is designed to work with the supply voltage of 6 volts required by this circuit.

✔ **Battery pack for 4 AA batteries**

✔ **Edmund Scientific's (www.edsci.com) 24"-diameter parabolic reflector, part #3053876**

✔ **SPST toggle switch, used as the on/off switch**

✔ **830-pin breadboard**

✔ **Five 2-pin terminal blocks**

✔ **Knobs for the potentiometer**

✔ **Two phono jacks**

✔ **Two right-angle phone plugs**

We use right-angle plugs to avoid having a loop of wire coming out of the box. (You can also use something called a banana plug and jack.)

✔ **Headphones**

We used a set of Philips HP170, but you can use any headphone you have available. Obviously, the better quality headphone you use, the better the sound quality.

✔ **Headphone jack**

We used a ¼" jack. If your headphone plug is a different size, use the appropriate size jack or get an adaptor.

✔ **Enclosure to protect the circuit**

We used a plastic box, RadioShack part #2701807.

✔ **An assortment of different lengths of prestripped, short 22 AWG wire**

✔ **PVC adaptor with ¾" female slip fitting on one end and a 1" thread fitting on the other end**

✔ **PVC 90° joint with a 1" slip fitting on both ends (one male and one female)**

✔ **PVC adaptor 1" female slip fitting to a 1" female thread**

✔ **PVC 1" end cap with a 1" female slip fitting**

✔ **⁵⁄₁₆"-diameter wooden dowel**

✔ Washer with a ¼" inner diameter (ID) and 1½" outer diameter (OD)

✔ ³⁄₁₆" ID rubber 0-ring

✔ ¼" thick rubber gasket material

✔ 1" diameter schedule 40 PVC pipe, 16" long

✔ 1" clamp

✔ Four ½" 8-32 panhead screws

✔ Four 8-32 nuts

You should be able to find the last 12 parts on this list at your local building supply or hardware store. Take this book with you: The photos should help you pick out the right parts, and everybody who sees you with it will realize just how smart you are.

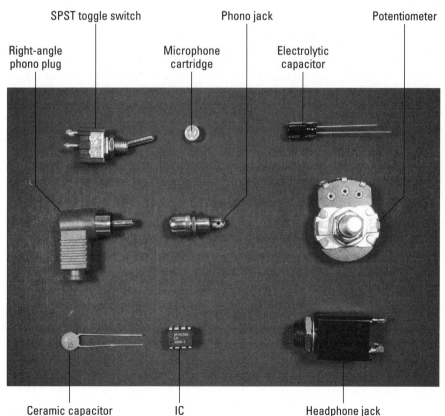

Figure 6-3:
Key
components.

SPST toggle switch — Phono jack — Potentiometer

Right-angle phono plug — Microphone cartridge — Electrolytic capacitor

Ceramic capacitor — IC — Headphone jack

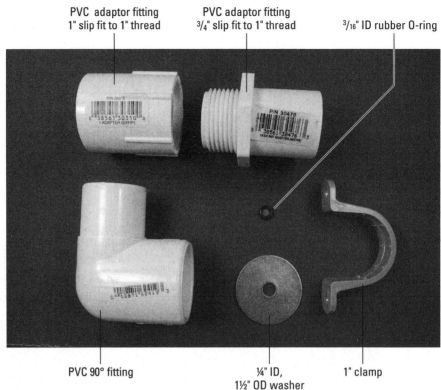

PVC adaptor fitting
1" slip fit to 1" thread

PVC adaptor fitting
³/₄" slip fit to 1" thread

³/₁₆" ID rubber O-ring

Figure 6-4:
More key
components.

PVC 90° fitting

¼" ID,
1½" OD washer

1" clamp

Taking Things Step by Step

We've taken a lot of the pain out of this project for you by finding just the right microphone, the right dish size, headphones, and so on so that you can easily pick up distant sounds. This took many mornings of standing on either end of our street shouting, "Say that again; I couldn't hear you!" Our neighbors still wonder about us, but it was worth it.

We further simplify your life by breaking down the process into a few easy steps.

Building an amplifier circuit

The first step in building a parabolic microphone is to tackle the circuit that forms its electronic brains. Here are the steps involved:

1. **Place LM386N-1 (IC2) and five terminal blocks (TB) on the breadboard, as shown in Figure 6-5.**

   The five terminal blocks shown in this figure will be used to connect two wires each to various components in the circuit. The wires from these five terminal blocks will go to the battery pack, the on/off switch, the microphone, the speaker, and the potentiometer.

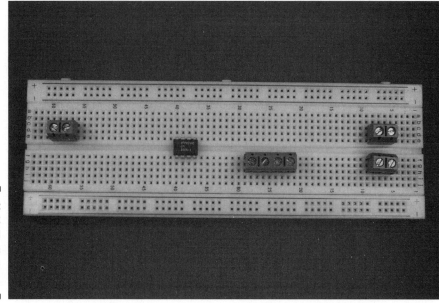

Figure 6-5:
Place the IC and terminal blocks on the breadboard.

2. **Insert wires to connect the IC and the terminal blocks to the ground bus and insert a wire between the two ground buses to connect them, as shown in Figure 6-6.**

   Six shorter wires connect components to ground bus; the long wire on the right connects the two ground buses.

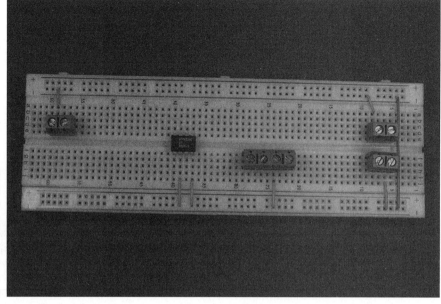

Figure 6-6:
Connecting
wires to the
IC, terminal
blocks, and
ground
buses.

3. **Insert wires to connect the IC and the terminal blocks to +V, and a wire between the two +V buses to connect them, as shown in Figure 6-7.**

Pin 6 of IC1 to +V   Battery TB to +V

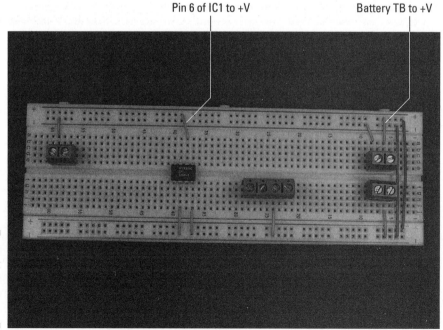

Figure 6-7:
Connect
components
to the
+V bus.

4. **Insert wires to connect the ICs, terminal block for the microphone cartridge, terminal blocks for the potentiometer (R2), terminal block for the headphone jack, and discrete components, as shown in Figure 6-8.**

Open region
to headphone jack TB

Pin 5 of IC1
to open region

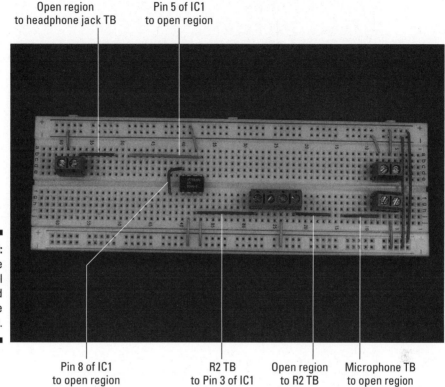

Figure 6-8:
Hook up the
ICs, terminal
blocks, and
discrete
components.

Pin 8 of IC1
to open region

R2 TB
to Pin 3 of IC1

Open region
to R2 TB

Microphone TB
to open region

5. **Insert the 0.047 microfarad capacitor (C5), two 10 microfarad capacitors (C2 and C3), one 100 microfarad capacitor (C4), one 0.1 microfarad capacitor (C1), one 5.6 kohm resistor (R1), and one 10 ohm resistor (R3) on the breadboard, as shown in Figure 6-9.**

When inserting electrolytic capacitors, be sure to check the schematic for where to insert the longer + lead.

We discuss in Chapter 4 how you can determine how short to clip the leads of many of these components to make them fit neatly on the breadboard. And you knew we were going to say it: Make sure you wear your safety glasses when clipping leads!

C3 from Pin 7
of IC1
to ground

C5 from Pin 5
of IC1
to open region

R3
from C5
to ground

C1 from
microphone TB
to R2 TB

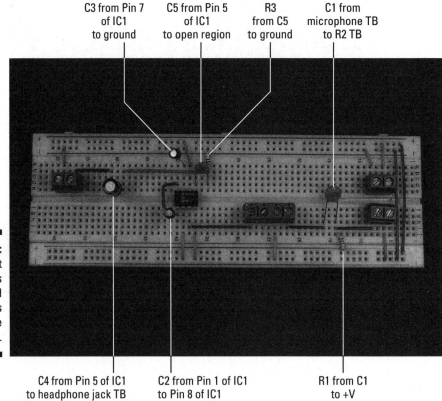

Figure 6-9:
Insert
resistors
and
capacitors
on the
breadboard.

C4 from Pin 5 of IC1
to headphone jack TB

C2 from Pin 1 of IC1
to Pin 8 of IC1

R1 from C1
to +V

Mounting everything on the dish

Time to assemble the handle that allows you to hold the parabolic microphone. Figure 6-10 shows how the pieces (minus the parabolic dish itself)
go together. When these fittings are assembled on the parabolic dish, the
threads go through the hole in the center of the dish with one gasket inside
the dish and one gasket outside the dish.

Gaskets   1" thread x 1" slip fitting

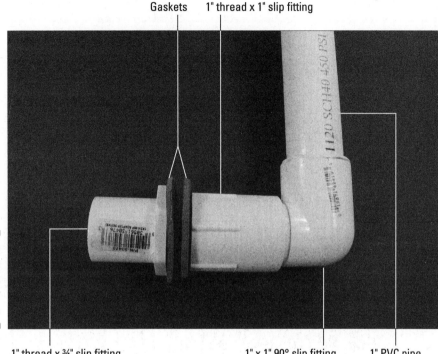

Figure 6-10:
Put the
fittings
together
like this.

1" thread x ¾" slip fitting   1" x 1" 90° slip fitting   1" PVC pipe

1. **In the sheet of gasket material, cut a hole (using your utility knife or X-ACTO knife) large enough to slip over the 1" threads.**

   Use one of the PVC fittings as a template.

2. **Use scissors to cut the outer diameter of the gasket about ⅜" larger than the inner diameter.**

3. **Repeat Steps 1 and 2 to create a second gasket.**

   Whenever you cut with a utility knife or an X-ACTO knife, wear leather work gloves to reduce the chance of cutting yourself if the knife slips.

4. **Glue a 16", 1"-diameter PVC pipe into the female end of the 90° fitting.**

5. **Glue the male end of the 90° fitting into the slip fit end of the 1" x 1" fitting.**

TIP

6. **Drill a ⅜" hole about eight inches from the end in the side of the PVC pipe that will be to your left when you're holding the parabolic microphone.**

   You use this hole to feed the wires from the microphone cartridge to the box containing your circuit.

7. **Slip one of the gaskets over the threads on the 1" x ¾" fitting. Then slip the threads from the inside through the hole in the dish.**

8. **Slip the other gasket over the threads and then screw the 1" x 1" fitting onto the threads.**

9. **Tighten the fitting by hand so that the gaskets are compressed a little and the dish is held securely between the gaskets.**

   Figure 6-11 shows the handle assembled on the dish.

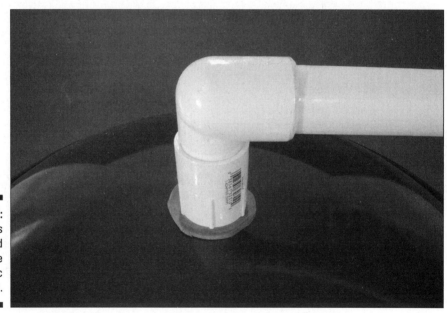

Figure 6-11:
Fittings
assembled
on the
parabolic
dish.

Mounting the microphone

The microphone is what makes this project work. The first step here is to make the microphone mounting. Follow these steps:

1. **Cut four 6" pieces of wooden dowel.**

2. **Cut a slot about one-third of the way through each dowel, ¼" from an end.**

   The slot has to be wide enough for the washer to slip into.

3. **Assemble the dowels and washer in the parabolic dish, as shown in Figure 6-12.**

   Use cable ties to hold the dowels in place. If you are doing this by your-self, it can be kind of a struggle, but it is doable. Getting another person to hold the dowels in place while you put the cable ties on it will simplify your life.

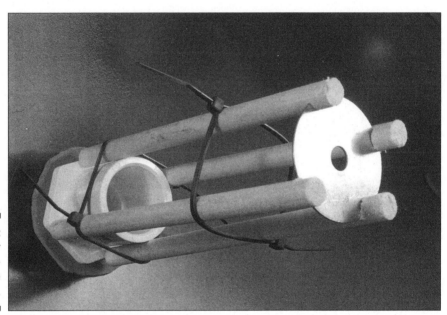

Figure 6-12: Put together the micro-phone mounting.

4. **Solder an 18" black wire to the ground pad of the microphone car-tridge and an 18" red wire to the +V pad.**

   Figure 6-13 shows a cartridge before and after soldering.

Black wire   Red wire

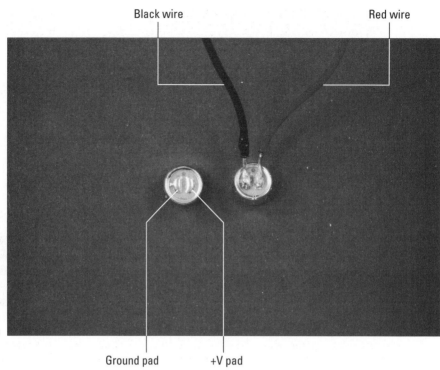

Figure 6-13:
Solder wires
to the
microphone
cartridge.

Ground pad   +V pad

5. **Twist the free ends of the wires from the microphone cartridge together and feed the microphone wires through the PVC fittings from inside the parabolic dish until the end of the wire reaches the ⅜" hole.**

If you have trouble pushing the wires around the 90° turn, take a stiff wire with a hook shape in the end (we used some 10 gauge wire; many building supply stores will sell you a few feet of this) and pull the wires through the PVC fittings and pipe.

6. **Use a piece of 20 or 22 gauge wire with a hook shape on the end to pull the wires from the microphone cartridge through the ⅜" hole.**

7. **Pull the wires through the ⅜" hole, leaving enough wire inside the parabolic dish for the microphone cartridge to reach the 1½" washer with about two inches of slack left over.**

8. **Cut the wires to allow three inches to extend from the ⅜" hole in the pipe and attach each wire to a right-angle phono plug, as shown in Figure 6-14.**

You can use either a plug that requires soldering to the wire or one that uses a screw to secure the wire, as we did here.

Figure 6-14:
Thread wires from the microphone cartridge through the PVC fittings and pipe.

9. **From the inside of the dish, put the ⅜" inner diameter O-ring over the microphone cartridge and glue the O-ring to the washer, as shown in Figure 6-15.**

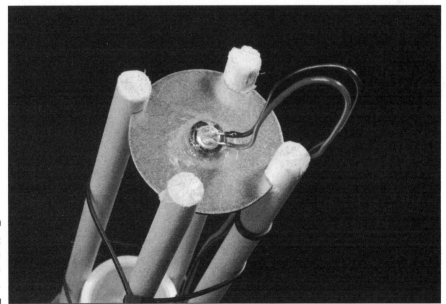

Figure 6-15:
Microphone cartridge in place.

Make sure you don't get glue on the front side of the microphone cartridge, or it could interfere with any sound coming through.

10. **Use another cable tie to secure the wires from the microphone to one of the dowels.**

Mounting switches and more on the box

It's time to drill all kinds of holes and pop various components into the box. Follow these steps to do so:

1. **Drill holes in the box where you will mount the potentiometer, audio jacks, headphone jack, clamp, and on/off switch, as shown in Figure 6-16.**

   We put the on/off switch and potentiometer on one side of the box, the headphone jack on the opposite side of the box, and the phono jacks on the bottom of the box.

   We kept the headphone jack to one side to keep the output signal from feeding into the input of the circuit (for example, by being near the microphone wires), which could cause feedback.

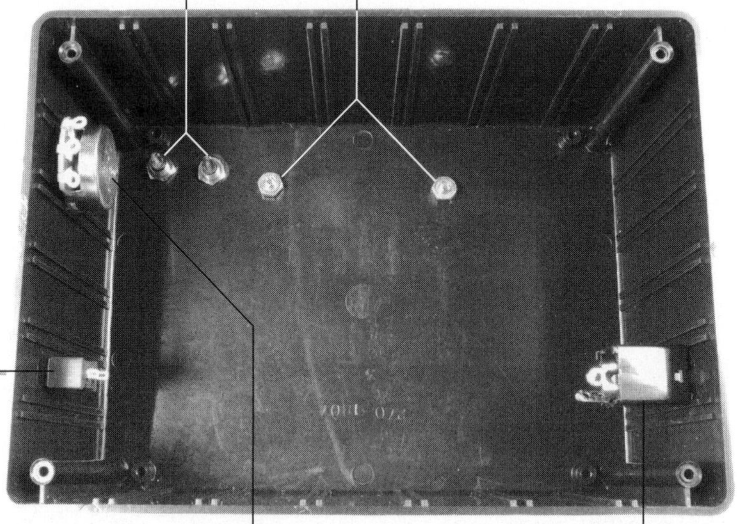

Screws and nuts
securing the clamp

Audio jack

Figure 6-16:
Box with
on/off
switch,
poten-
tiometer,
headphone
jack, clamp,
and audio
jacks.

On/Off switch   Potentiometer   Headphone jack

See Chapter 4 for more information about choosing drill bit sizes for particular components and other advice on how to customize a box for your projects. Make sure you use safety glasses when drilling and clamp the box to your worktable!

2. **Slip the shaft of the on/off switch through the drilled hole and secure with the nut provided.**

3. **Slip the shaft of the potentiometer through the drilled hole and secure with the nut provided.**

4. **Slip the knob on the potentiometer shaft and secure with the set screw.**

5. **Slip the threads of the headphone jack through the drilled hole and secure with the nut provided.**

6. **Slip the threads of the audio jacks through the drilled holes and secure with the (you guessed it) nuts provided.**

7. **Slip 8-32 screws through holes in the clamp and the holes you drilled in the box and secure with nuts from the inside of the box.**

8. **Solder the black wire from the battery pack to one lug of the on/off switch and solder an 8" black wire to the remaining lug of the on/off switch, as shown in Figure 6-17.**

9. **Solder one 8" black wire to the lug on one of the audio jacks and solder one 8" red wire to the lug on the other audio jack, as shown Figure 6-17.**

10. **Solder 8" wires to each of the three potentiometer lugs, as shown in Figure 6-17.**

11. **Solder one 8" black wire and one 8" red wire to the lugs of the headphone jack, as shown Figure 6-17.**

The red wire should go to the contact that touches the tip of the headphone plug.

Okay, we'll say it again: Heed all the safety precautions about soldering that we give you in Chapter 2. Don't ever leave your soldering iron on and unattended. And to avoid damage from flying pieces of solder, wear your safety glasses!

Wires to
potentiometer
lugs

Black wire to phono jack

Red wire to phono jack

Red wire
to headphone
jack

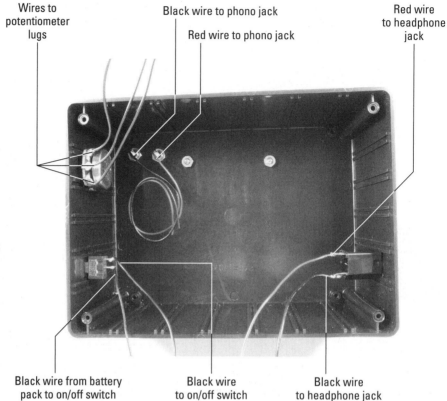

Figure 6-17:
Solder wires
to the on/
off switch,
phono plugs,
potentiome-
ter, and
headphone
plug.

Black wire from battery
pack to on/off switch

Black wire
to on/off switch

Black wire
to headphone jack

Putting everything together

After you have a completed breadboard, all the switches in the box, and the microphone and dish assembled, it's time to put all those elements together.

Follow these steps to finish building your parabolic microphone:

1. **Attach Velcro to the breadboard and the box and secure the bread-board in the box.**

2. **Attach Velcro to the battery pack and the box and secure the battery pack in the box.**

3. **Insert the wires from the headphone jack, potentiometer, phono jacks, battery pack, and the on/off switch to the terminal blocks on the breadboard, as shown in Figure 6-18.**

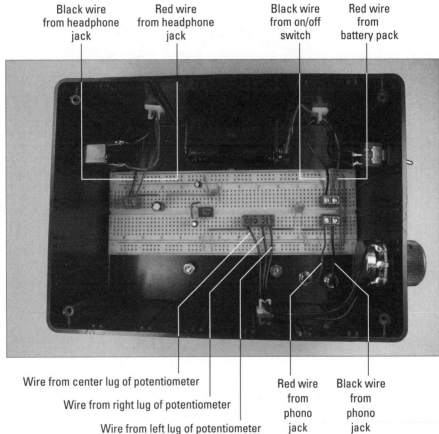

Black wire from headphone jack

Red wire from headphone jack

Black wire from on/off switch

Red wire from battery pack

Figure 6-18: Connect the headphone jack, potentiometer, on/off switch, and battery pack to the breadboard.

Wire from center lug of potentiometer

Wire from right lug of potentiometer

Wire from left lug of potentiometer

Red wire from phono jack

Black wire from phono jack

4. **As you insert the wires, cut each of them to the length you need to reach the assigned terminal block and strip the insulation from the end of each wire.**

Keep the wires from the headphone jack as far as possible away from the wires from the potentiometer and the wires from the microphone cartridge. Remember how a microphone put too close to a speaker can produce an awful screech? The same screech can assault your ears if these wires get too close together.

5. **Secure the wires with wire clips where needed.**

   The way that parts are laid out in this box as well as the distance between some of the components — such as from the phono jacks to the terminal blocks — are both short enough that you won't need wire clips.

6. **Slip the box onto the PVC pipe and tighten the clamp screws.**

   Don't tighten them too much; you still need to slide the box into its final position.

7. **Put batteries in the battery pack and put the lid on the box, securing it with the screws provided with the box.**

8. **Slide the box into its final position just below the hole where the phono jacks come out of the PVC pipe. Then insert the phono plugs and headphone plug into the jacks in the box, as shown in Figure 6-19.**

Figure 6-19: The electronics in place.

9. **Glue the 1" PVC end cap on the end of the PVC pipe.**

   You can see the finished product at the beginning of the chapter in Figure 6-1.

Trying It Out

At this point, the microphone is ready to go, but here are a few tips to help you be kind to your ears — you can get some really loud noises out of this thing.

> ✔ ***Before putting on the headphones,*** **flip the switch on and adjust the volume control.**
>
> You'll be able to hear enough to make sure you don't have the volume so high that it hurts your ears.

> ✔ **Avoid knocking the parabolic dish against anything like a tree trunk or your cat while you have the headphones on.** The resultant ringing in your ears is not a desirable thing.

> ✔ **Avoid having any "friends" shout into the parabolic dish, or anywhere near it, while you have the headphone on.**

Time to get this thing in gear. Here goes:

With the headphones on your head, flip the On switch, point the thing at something you want to listen to (such as a neighborhood bird or your best friend), and listen.

If you don't get the results we got, here are some options to check out:

> ✔ Check that all the batteries are fresh, tightly inserted in the battery pack, and face the right direction.

> ✔ Check that no wires or components have come loose.

> ✔ Compare your breadboard with the photos to make sure all the wires and components are connected correctly.

Taking It Further

After you wander around your neighborhood with your parabolic microphone (and explain to curious passersby what that thing is), you might want to try some variations. Here are our suggestions:

> ✔ Instead of headphones, hook up a tape recorder to the circuit and record sounds. (Once again, do not record your neighbors because there is probably some kind of law against this — and if there isn't, there should be.)

- ✔ Get a bigger dish and set it up outside, hooked into speakers you set up in a permanent wildlife listening station either outdoors or in your house. (We say *permanent* because a dish much bigger than 24 inches needs a wheelbarrow to cart it around.)

- ✔ If you replace the microphone with a speaker, you could turn this thing on its head and make it a loudspeaker instead of a long-distance microphone. Look up any noise ordinances on the books before trying this one!

Chapter 7

Murmuring Merlin

A picture can be worth a thousand words, but sometimes a word — or sound — is just what's needed. In this project, you work with a sound synthesizer chip, an amplifier, and a speaker. With this setup, you can produce almost any sound you can imagine.

You can place a sound chip in just about anything and use switches to activate sounds. For this project, we chose a hand puppet (ours happens to be a wizard because we're into fantasy fiction); a hollow puppet lets you easily insert the project breadboard and switches. However, you can use anything you like for your talking pal.

As you work through this project, you discover how to work with programmable sound synthesizer chips as well as a bit about how amplification works.

The Big Picture: Project Overview

When you complete this project, you'll have a talking hand puppet. You can program the synthesizer chip inside the puppet with any messages you like. For example, we programmed ours to say, "The check's in the mail," "You can't have more money," and "Where's the darn chapter?!" — and then we gave it to our editor.

These messages (or sounds) are activated when you shake either of the puppet's hands or press the puppet's nose. Alas, adding the electronics to the puppet makes it unsuitable for use on your hand, but you can't have everything.

You can see the finished talking puppet (okay, you have to imagine the sound part) in Figure 7-1.

Figure 7-1:
We affectionately call our talking puppet *Murmuring Merlin.*

Here are the types of activities that you'll do to create your own talking toy:

1. Put together an electronic circuit containing a sound synthesizer chip, an amplifier, and a speaker.

2. Hook up the circuit to your computer to program the sound synthesizer with sentences, music, or whatever strange sounds you'd like to play around with.

3. Place the box containing the circuit in the puppet and connect switches in the puppet to the circuit.

After these steps, when someone presses either of the puppet's hands or presses its nose, it plays whatever you programmed in the sound synthesizer for that particular switch.

 Although the hand puppet we work with in this chapter might seem like a toy, it's not intended for small children. The wires and small electronic components could be swallowed, and you don't certainly want small kids playing with batteries.

Scoping Out the Schematic

You have only one breadboard to put together for this project. You can see the schematic for the board in Figure 7-2.

The following is a list of the schematic elements for our Murmuring Merlin.

- ✔ **IC1** is a SpeakJet sound synthesizer model. You can connect this IC to your computer and program it to generate electrical signals by using software supplied by the manufacturer. These signals correspond to words that form sentences, sounds that create music, or various cool sound effects. The software allows you to program eight sounds; each sound is controlled by one of the Pins 1–4 and 6–9 of the IC.

- ✔ **C1** is a capacitor that filters noise from the +4.5V supply to IC2.

- ✔ **Switches S1, S2, and S3** control the voltage on Pins 2, 4, and 7. The switches are normally *open,* which means each pin is normally connected to ground. When you push one of the switches, the voltage on the corresponding pin raises to +4.5 volts. When you release the switch, the voltage on the pin returns to ground. Because we programmed the SpeakJet to trigger when the voltage on a pin changes from high (+4.5 volts) to low (ground), the sound which that pin controls is triggered when you press and release the corresponding switch. We used three switches to control three sounds because the puppet we used has three handy spots where we could place switches. (If you're lucky enough to find an octopus puppet, you can have up to eight switches, controlling up to eight sounds.)

- ✔ **IC2,** a MAX232 driver/receiver chip, converts the signals from your computer. These signals aren't generated with the correct voltage for this circuit, so signals have to be converted so that they can be used by the SpeakJet chip. This chip also converts signals from the SpeakJet into signals that your computer can use.

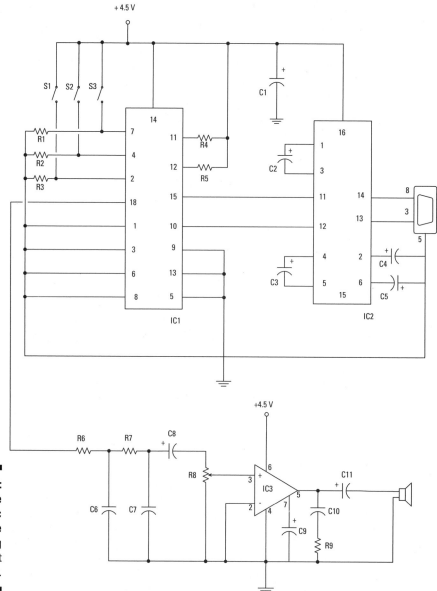

Figure 7-2:
The
schematic
of the
talking
puppet
circuit.

✔ **Resistors R1, R2, and R3** limit the current running to ground when you press switches S1, S2, or S3.

✔ **C2, C3, C4, and C5** are capacitors that fill a function that the Max232 designers call *charge pump capacitors*. These are required to make the IC2 function properly.

- **R6 and R7, C6 and C7** are two resistors and two capacitors, respectively, that form a filter (a *low pass* filter) that eliminates high-frequency noise prior to the signal reaching the amplifier.

- **C8** is a capacitor that removes any DC offset from the output of IC1.

- **R8** is a potentiometer that controls the sound volume.

- **IC3,** an LM386N-1 audio amplifier, takes the electrical signal generated by the SpeakJet when you push and release one of the switches in the puppet and then amplifies the electrical signal to create enough power to drive the speaker.

- **C9** is a capacitor that improves the stability of the LM386 amplifier to prevent problems such as oscillation.

- **Capacitor C10** acts as a current bank for the output. This capacitor drains when sudden surges of current occur and refills with electrons when the demand for current is low.

- **C11** is a capacitor that removes any DC offset from the output of the LM386 amplifier.

Building Alert: Construction Issues

The tactile switches that we use in this project have very tiny leads that are meant to be surface mounted in an assembly line. Given that we (and you) don't have an assembly line handy, we used needlenose pliers to crimp wire around the tiny leads to hold the wires in place while we soldered them.

The DB9 connector has a small metal tube or cup to which you can solder each wire connection. We found that the easiest way to solder a wire to one of these tubes is to melt some solder into the tube, reheat the tube, and then insert the bare wire end into the melted solder.

When inserting electrolytic or tantalum capacitors into the breadboard, pay attention to the polarity. Inserting the capacitors the wrong way could damage the capacitors and possibly other components in your circuit. The longer lead of the capacitor is the + side. The schematic shows you which direction to insert the capacitor. For example, the + lead of capacitor C2 goes toward Pin 1 of IC2, and the other lead goes toward Pin 3 of IC2.

In order to feed wires from the tactile switches to a location where they could be connected to the electronics box by the shortest path, we had to cut a few holes in the fabric of the puppet. Don't worry; he won't feel a thing.

Perusing the Parts List

Your biggest shopping decision for this project is what to place the project in. The hand puppet idea is nice because it's hollow and has a personality, as opposed to using just a plain wooden box. However, you could technically put the insides of the project into anything. For example, you could cut open a stuffed toy, take out the stuffing, and put the project inside.

Assuming you take our route and go with a hand puppet, though, pick out one that you like with enough room to fit the electronics enclosure and with openings that allow you to place switches in the hands, face, or other areas. We found one at a local toy store made by Sunny and Co. Toys, Inc., their item #GL1902.

After you choose your housing unit, shop for the parts for the project itself, several of which are shown in the upcoming Figures 7-3 and 7-4:

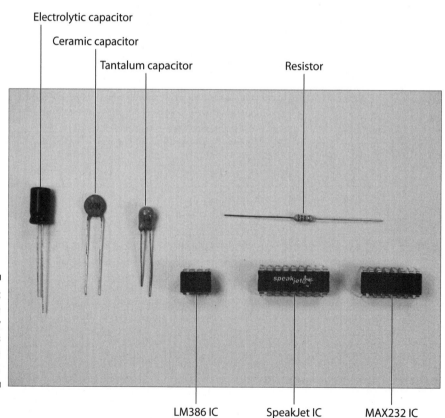

Figure 7-3: Many of the key components for the project.

Electrolytic capacitor

Ceramic capacitor

Tantalum capacitor

Resistor

LM386 IC   SpeakJet IC   MAX232 IC

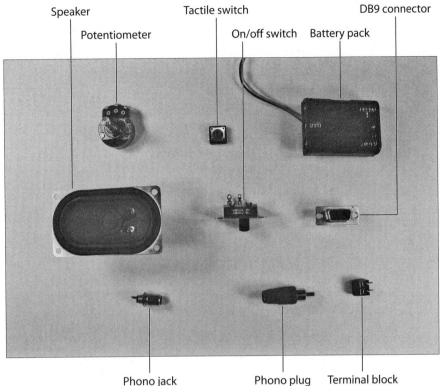

Speaker

Potentiometer

Tactile switch

On/off switch   Battery pack

DB9 connector

Figure 7-4:
Key
components,
part 2!

Phono jack   Phono plug   Terminal block

- ✔ **10 kohm potentiometer (R8)**

- ✔ **Two 33 kohm resistors (R6 and R7)**

- ✔ **Five 1 kohm resistors (R1, R2, R3, R4, R5)**

- ✔ **10 ohm resistor (R9)**

- ✔ **0.01 microfarad ceramic capacitor (C6 and C7)**

- ✔ **Five 1 microfarad tantalum or electrolytic capacitor (C1, C2, C3, C4, C5)**

- ✔ **0.047 microfarad ceramic capacitor (C10)**

- ✔ **Three 10 microfarad electrolytic capacitors (C8 and C9)**

- ✔ **100 microfarad electrolytic capacitor (C11)**

- ✔ **Battery pack for three AAA batteries**

We use three AAA batteries in this project to supply 4.5 volts. If we used a four-battery pack as we do in other chapters, the supply voltage would be about 6 volts, above the maximum supply voltage allowed for the SpeakJet IC.

✔ **SpeakJet sound synthesizer IC1**

You can find a list of distributors on the manufacturer's Web site, www.speakjet.com.

✔ **LM386N-1 amplifier IC3**

Of the many versions of the LM386 amplifier, we chose the LM386N-1 because it works with the supply voltage of 4.5 volts used by this circuit.

✔ **MAX232 driver/receiver IC2**

✔ **SPST (single-pole, single-throw) slide switch, used as the on/off switch**

✔ **830-contact breadboard**

✔ **Eight 2-pin terminal blocks**

✔ **Knob (for potentiometer)**

✔ **8 ohm, 1 watt speaker**

✔ **Three tactile switches (S1, S2, S3)**

Many of these tactical switches are very small; for example, a 6 mm × 6 mm switch is less than ¼" square. We bought 12 mm × 12 mm switches (part #TS6424T2602AC) made by Mountain Switch at Mouser (www.mouser.com).

✔ **Enclosure**

We used Radio Shack part #270-10807.

✔ **Six phono plugs**

✔ **Six phono jacks**

✔ **DB9 female connector**

There are about as many DB9 connecters out there as there are fish in the sea. We used RadioShack part #276-1538.

✔ **DB9 serial port cable**

✔ **Four ½" 6-32 flathead screws**

✔ **Four 6-32 nuts**

✔ **An assortment of different lengths of prestripped, short 22 AWG wire**

Taking Things Step by Step

If we were wizards like Merlin, we could probably snap our fingers, say "Abra-cadabra," and produce a talking puppet. Being simple electronic project geeks, though, we (and you) have to do things the old-fashioned way. Building a talking puppet requires that you

1. Create the circuit.

2. Drill various holes in a box.

3. Place the circuit in the box.

4. Program the sound synthesizer.

5. Insert the box and switches into the puppet.

Creating Merlin's circuit

Get your breadboard and other little parts ready to build the circuit that will help Merlin murmur. Here are the steps involved:

1. **Place the SpeakJet IC (IC1), the MAX232 IC (IC2), the LM386 IC (IC3), and eight terminal blocks on a breadboard, as shown in Figure 7-5.**

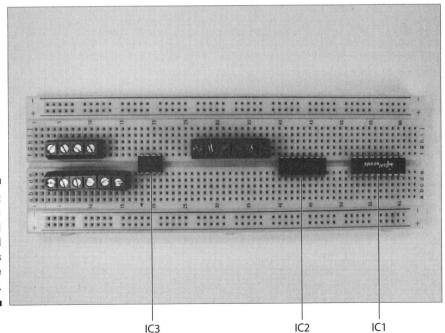

Figure 7-5:
Place the ICs and terminal blocks on the breadboard.

IC3   IC2   IC1

The eight terminal blocks shown in this figure will be used to connect two wires each to various components in the circuit. The wires from these terminal blocks go to the battery pack, the DB9 connector, the on/off switch, tactile switches in the puppet, the speaker, and the potentiometer, respectively.

2. **Insert wires to connect the ICs and the terminal blocks to the ground bus and then insert a wire between the two ground buses to connect them to each other, as shown in Figure 7-6.**

   Fourteen shorter wires connect components to ground bus; the long wire on the left connects the two ground buses.

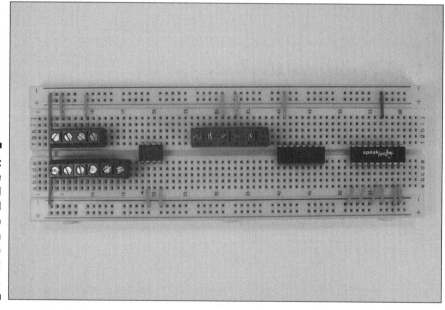

Figure 7-6:
Connect the ICs and terminal blocks to ground; then connect the ground buses.

3. **Insert wires to connect the ICs and the terminal blocks for the battery pack and the three tactile switches (S1, S2, S3) to the +V bus, as shown in Figure 7-7.**

4. **Insert wires to connect the ICs; terminal blocks for the potentiometer (R8) and tactile switches (S1, S2, S3); and discrete components, as shown in Figure 7-8.**

5. **Insert discrete components on the breadboard, as shown in Figure 7-9.**

   When inserting electrolytic or tantalum capacitors, be sure to check the schematic to see where to insert the longer, positive (+) lead. The components shown in this figure, indicated by numbered callout, include

Battery terminal block to +V

Pin 6 of IC3 to +V

Pin 16 of IC2 to +V

Pin 14 of IC1 to +V

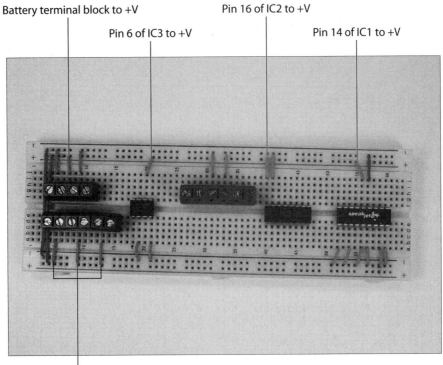

Figure 7-7:
Connect
components
to the +V
bus.

Tactile switch terminal blocks to +V

1. 1 kohm resistor R1 from IC1 Pin 2 to ground

2. 1 kohm resistor R2 from IC1 Pin 4 to ground

3. 1 kohm resistor R3 from IC1 Pin 7 to ground

4. 1 kohm resistor R4 from IC1 Pin 11 to +V

5. 1 kohm resistor R5 from IC1 Pin 12 to +V

6. 1 microfarad capacitor C2 from IC2 Pin 1 to Pin 3

7. 1 microfarad capacitor C3 from IC2 Pin 4 to Pin 5

8. 1 microfarad capacitor C4 from IC2 Pin 2 to ground

9. 1 microfarad capacitor C5 from IC2 Pin 6 to ground

10. 1 microfarad capacitor C1 from +V to ground

11. 33 kohm resistor R6 from wire to IC1 Pin 18 to an open region

12. 33 kohm resistor R7 from R6 to an open region

13. 10 microfarad capacitor C8 from R7 to wire to R8 terminal block

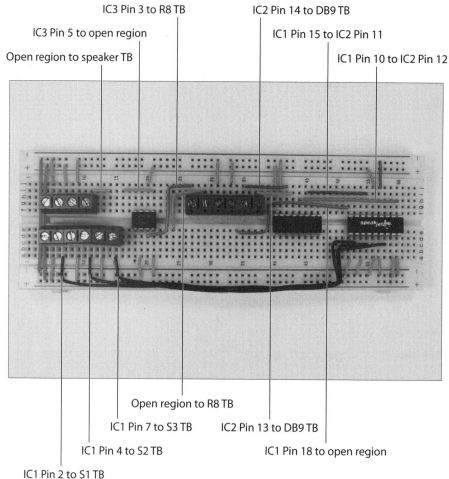

IC3 Pin 3 to R8 TB

IC3 Pin 5 to open region

Open region to speaker TB

IC2 Pin 14 to DB9 TB

IC1 Pin 15 to IC2 Pin 11

IC1 Pin 10 to IC2 Pin 12

Figure 7-8:
Hook up the
ICs, terminal
blocks (TB),
and discrete
components.

Open region to R8 TB

IC1 Pin 7 to S3 TB

IC1 Pin 4 to S2 TB

IC2 Pin 13 to DB9 TB

IC1 Pin 18 to open region

IC1 Pin 2 to S1 TB

14. 0.01 microfarad capacitor C6 from R6 to ground

15. 0.01 microfarad capacitor C7 from R7 to ground

16. 10 microfarad capacitor C9 from IC3 Pin 7 to ground

17. 0.047 microfarad capacitor C10 from IC3 Pin 5 to open region

18. 10 ohm resistor R9 from C10 to ground

19. 100 microfarad capacitor C11 from wire to IC3 Pin5 to wire to speaker terminal block

We discuss in Chapter 4 how you can determine how short to clip the leads of many of these components to make them fit neatly on the breadboard.

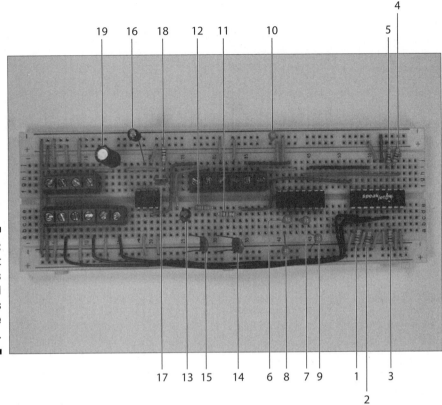

19 16 18 12 11   10   5   4

Figure 7-9:
Insert
resistors
and
capacitors
on the
breadboard.

17 13 15   14   6 8 7 9   1   3
2

Make sure you wear your safety glasses when clipping leads!

6. **Solder 12" wires to the solder cups for Pins 3, 5, and 8 of the DB9 connector.**

The location of these pins is shown in Figure 7-10. Although the manufacturer prints the pin numbers next to the solder cups, you might need a magnifying glass to read the pin numbers on the connector.

Be sure to heed all the safety precautions about soldering that we list in Chapter 2. For example, don't leave your soldering iron on and unattended. And just in case a bit of solder has an air pocket and pops, wear your safety glasses!

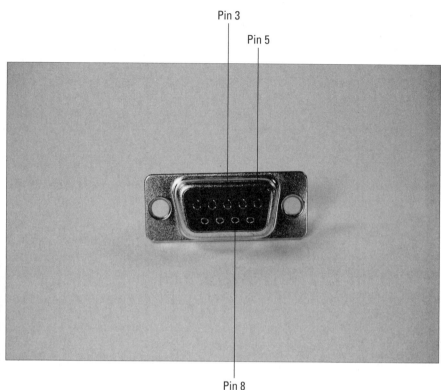

Pin 3

Pin 5

Figure 7-10:
Solder cups
on back of
the DB9
connector.

Pin 8

7. **Slip a 1" length of heat shrink tubing along each wire until it covers the soldered joints. Then heat the tubing with a hair dryer to secure it and protect the joints.**

Making the box puppet-friendly

It's time to prepare the box so that you can insert the guts of the project into it and string wires to the puppet. This involves some drilling, some cutting, and some assembling.

Follow these steps to get the project box wired up:

1. **Drill holes in the box for the phono jacks, potentiometer, speaker, DB9 connecter, and on/off switch in locations as shown in upcoming Figure 7-11.**

We used a ½" drill bit for the phono jacks and a 5/16" bit for the potentiometer. To secure the speaker, we drilled four holes using a 5/64" bit. You might need to use different bit sizes, depending on the parts you use.

Wear your safety glasses whenever you drill holes or cut wires. Also, the drill bit can bind as it goes through the plastic, causing the box to turn with the drill if it's not properly secured. Don't test Murphy's Law: Use a vise or other method to secure the box while you're drilling.

2. **Cut openings for the DB9 connector, on/off switch, and the speaker.**

   We drilled a pilot hole and then used a coping saw to cut the openings. The openings don't have to be the exact shape of the part. For the DB9 connector and the switch, make the openings big enough for the body of the part to fit through. The openings can even be a little oversized as long as enough plastic is left for you to secure the part's flange. For the speaker, cut an opening about ⅜" inside the outline of the speaker. All you need is an opening big enough to let the sound travel.

3. **Insert the phono jacks, potentiometer, speaker, DB9 connector, and on/off switch as shown in Figure 7-11.**

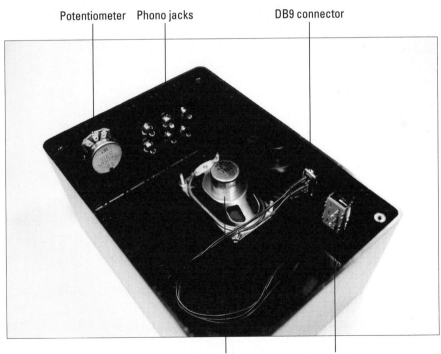

Potentiometer   Phono jacks   DB9 connector

Speaker   On/off switch

Figure 7-11:
Mount components on the box.

4. **Secure the phono jacks and potentiometer with the nuts provided with each part.**

   The outside of the box is shown in Figure 7-12.

   You can secure the speaker with four ½" 6-32 flathead screws and four 6-32 nuts. Use glue on the flanges of the DB9 connector and on the on/off switch to attach them to the box.

Figure 7-12:
A view from
the outside.

5. **Solder the black wire from the battery pack to one lug of the on/off switch and solder an 8" black wire to the other lug of the on/off switch, as shown in Figure 7-13.**

6. **Solder a 6" black wire to each of the two solder lugs on the speaker, as shown in Figure 7-13.**

7. **Solder an 8" wire to each of the three potentiometer lugs, as shown Figure 7-14.**

8. **Connect 8" wires to each of the audio jacks and solder them, as shown in Figure 7-14.**

9. **Use Velcro to attach the breadboard and the battery pack to the box.**

Black wire from battery pack to on/off switch

Black wire to on/off switch

Figure 7-13:
Wires
soldered to
the on/off
switch and
speaker.

Wires to speaker

10. **Cut the wires from the phono jacks, potentiometer, on/off switch, battery pack, and DB9 connector to a length that allows you to arrange wires neatly in the enclosure.**

11. **Strip insulation from the ends of cut the wires, insert them in the terminal blocks, and secure with wire clips, as shown in Figure 7-15.**

    The components shown in this figure, indicated by numbered callout, include

    1. Wires from left phono jack

    2. Wires from center phono jack

    3. Wires from right phono jack

    4. Wire from left potentiometer lug

    5. Wire from center potentiometer lug

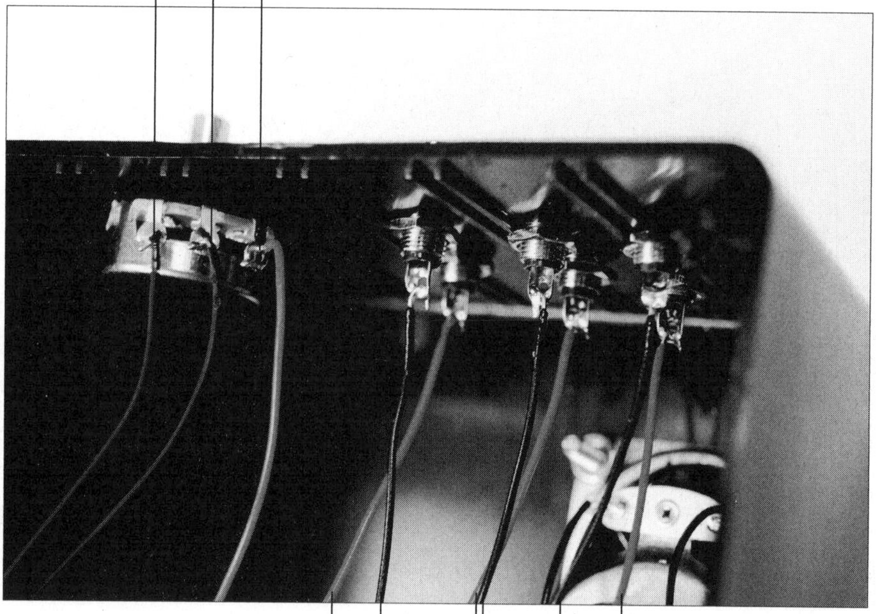

Wire soldered to left potentiometer lug

Wire soldered to center potentiometer lug

Wire soldered to right potentiometer lug

Figure 7-14:
Solder wires
to connect
external
components
to the
breadboard.

Wires soldered to audio jacks

6. Wire from right potentiometer lug

7. Wire from Pin 3 of DB9 connector

8. Wire from Pin 5 of DB9 connector

9. Wire from Pin 8 of DB9 connector

10. Red wire from battery pack

11. Black wire from on/off switch

12. Speaker wires

**12. Place the on/off switch in the off position, insert batteries in the
battery pack, and secure the lid on the enclosure with the screws
provided.**

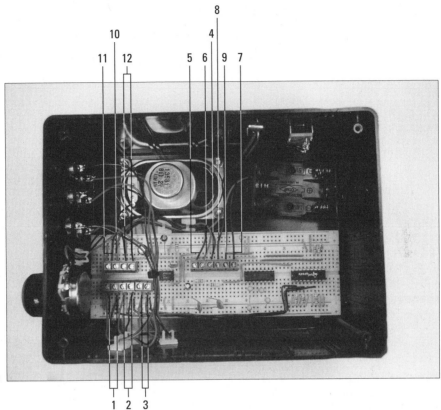

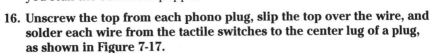

Figure 7-15:
Insert wires
into terminal
blocks.

13. Use your multimeter to find two contact pins on the tactile switches that are normally *open* (infinite resistance between the contact pins) and *closed* (nearly zero resistance between the two pins) when the switch is pressed.

14. Solder a 12" wire to each of the two pins on each tactile switch, as shown in Figure 7-16.

15. Cut the wires from the tactile switches to the length you need to reach the jacks on the electronics enclosure after both the switches and the box are in place.

Leave about three extra inches of length to allow for some shifting as you stuff the box in the puppet.

16. Unscrew the top from each phono plug, slip the top over the wire, and solder each wire from the tactile switches to the center lug of a plug, as shown in Figure 7-17.

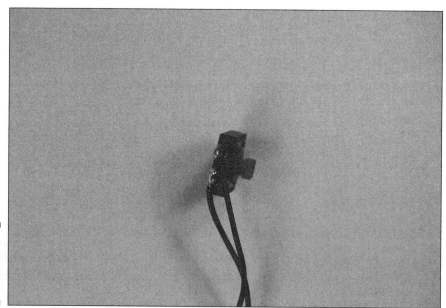

Figure 7-16:
Solder wires
to tactile
switches.

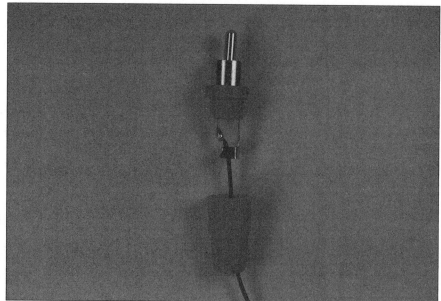

Figure 7-17:
Solder wires
to phono
plugs.

17. **Screw the covers of the phono plugs back in place and feed each tactile switch to the proper location.**

    We placed one switch in each hand and a switch under the puppet's nose.

18. **Feed the phono jacks behind the puppet, as shown in Figure 7-18.**

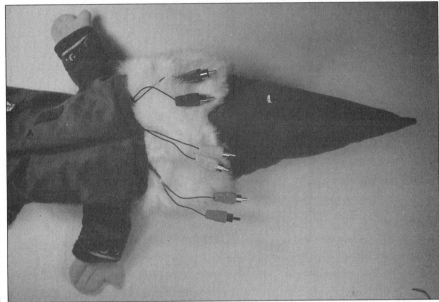

Figure 7-18:
Place tactile
switches in
the puppet.

You might want to secure the switches in place in some fashion. We used a strip of adhesive-backed Velcro (somewhat wider than the switch) for the switch behind the nose. We stuck this to the back of the switch, put the switch in place, and pressed the adhesive back to the fabric, right under the puppet's nose. Another idea is to use a few stitches of thread to stop the switch from moving.

Programming sounds

Even with your picture-perfect puppet and circuit, Merlin won't murmur a thing without sounds programmed into the sound synthesizer chip. For this part, you have to study a bit on the software provided by the manufacturer. SpeakJet doesn't provide a lot of documentation for using the software, but we felt that the chip was the best bet for our puppet because it offers so many cool options for creating sounds.

The steps that we provide here are just an introduction to programming the SpeakJet. The SpeakJet user guide is available at `www.speakjet.com`, under the Documentation heading.

If you need more information, we suggest you try the SpeakJet discussion forum at

```
http://groups.yahoo.com/group/speakjet
```

You can get the PhraseALator software that you use to program the chip at

```
http://magnevation.com/software/PhraseALator.zip
```

The manufacturer assures us that it is creating a manual for the PhraseALator. Go to `www.speakjet.com` and check under Documentation to see whether it's been posted.

By using this chip and software to program it, you can get some pretty neat functionality. For example, at your disposal is a built-in set of 72 speech elements, 43 sound effects, 3 octaves of musical notes, and 12 touch tones. By mixing and matching these and controlling the pitch, rate, bend, and volume settings, you can produce just about any sound, phrase, or musical tones you want.

After you install the software on your computer, you can simply assign pre-loaded sounds to the circuit of this project, which we cover in the next set of steps. Or, you can get fancy and start programming sounds or even words of your choosing. If you want to get fancy, it will take some playing around to learn the software (which is beyond the scope of this book). As we said, not much extensive documentation is available for it at this point, but with a little trial and error, you can get the hang of it.

For our purposes, we used the sounds that the manufacturer preloaded. Here are the steps involved in this procedure:

1. **Open the PhraseALator software and click the Event Configuration button.**

   The upcoming Figure 7-19 shows the SpeakJet Event Configuration that we loaded onto our chip. The items in the Phrase# to Play column in this figure indicate phrases that the manufacturer has preloaded.

2. **Select the check boxes in the Play Phrase column only for input events listed as Goes Low.**

You do this so that the SpeakJet activates when you first press and then release the switch. See the note and table after these steps for help in choosing which check boxes you must select. (The simple route is to just select all the Goes Low inputs, as shown in Figure 7-19.)

3. **Connect the box to your computer by using a serial port cable to program the SpeakJet.**

   Check over your circuit to make sure that no wires are loose or touch another wire — which could cause a short — before hooking them up to your computer.

4. **With the box connected to your computer, simply click the Write Data to SpeakJet button in the Event Configuration screen of the SpeakJet program (as shown in Figure 7-19).**

Figure 7-19:
The Event Configuration screen of the PhraseA-Lator software program.

	Play Phrase	Phrase# To Play	Clear Buffer	Call Phrase	Restart From Wait
SpeakJet Event Configuration					
Input #0 - Goes High	☐	0	☑	☐	☐
Input #0 - Goes Low	☑	10	☑	☐	☐
Input #1 - Goes High	☐	1	☑	☐	☐
Input #1 - Goes Low	☑	1	☑	☐	☐
Input #2 - Goes High	☐	2	☑	☐	☐
Input #2 - Goes Low	☑	2	☑	☐	☐
Input #3 - Goes High	☐	3	☑	☐	☐
Input #3 - Goes Low	☑	3	☑	☐	☐
Input #4 - Goes High	☐	4	☑	☐	☐
Input #4 - Goes Low	☑	4	☑	☐	☐
Input #5 - Goes High	☐	5	☑	☐	☐
Input #5 - Goes Low	☑	5	☑	☐	☐
Input #6 - Goes High	☐	6	☑	☐	☐
Input #6 - Goes Low	☑	6	☑	☐	☐
Input #7 - Goes High	☐	7	☑	☐	☐
Input #7 - Goes Low	☑	7	☑	☐	☐
Power up & Reset	☑	2			

Input Style

	TTL Signal	RC Pulse
Input #7	⦿	○
Input #6	⦿	○

Output Controled by

	Chip	Phrase
Output #0	⦿	○
Output #1	⦿	○
Output #2	⦿	○

Output Startup State

	On	Off
Output #0	⦿	○
Output #1	⦿	○
Output #2	⦿	○

Write Data to SpeakJet

Miscellaneous
☑ Auto Silence   SCP Node: 0   Shut Up   Done

5. **After you finish programming the chip, disconnect the box with the electronics from the computer.**

The event number in the SpeakJet software Event Configuration window doesn't always match the pin number in the SpeakJet IC — that would be too easy. Here's a list of event numbers and corresponding pin numbers. We wired switches to Pins 2, 4, and 7, so we use Events 6, 4, and 2.

Event	Pin
0	9
1	8
2	7
3	6
4	4
5	3
6	2
7	1

Hooking up the puppet

Merlin is just bursting to have his say at this point, so it's time to take the final steps to activate the newly loquacious puppet.

Making sounds

We know you're the inquiring kind who will want to program custom phrases before long, so here's an overview of the process to get you started:

1. **Open the EEPROM Editor screen.**

2. **Select the check box for the Phrase# that you wish to use for a custom sound.**

3. **In the Phrase Editor screen that opens, select the sounds that you wish to include in a phrase.**

   Be sure to check out words available from the library.

4. **After you make a custom phrase, click Done in the bottom right of the screen.**

This takes you back to the EEPROM Editor screen, where you can see that the selected sounds have been added to the phrase.

5. **Close the EEPROM Editor screen and open the Event Configuration screen.**

6. **Enter the phrase number that you created in the Phrase# to Play box for one of the input #s.**

   See the nearby table to see how input numbers correspond to pin numbers.

7. **Connect the serial cable and then click the Write Data to SpeakJet button to program the chip.**

Follow these steps to hook up and operate the puppet:

1. **Plug the phono plugs from one of the tactile switches into the box and press the switch.**

   If the sound is too soft or too loud, adjust the potentiometer to get the sound level right.

2. **Unplug the phono plugs and insert the electronics box in the puppet, taking care not to move the tactile switches out of position.**

3. **Insert each phono plug in a phono jack and tuck the wires out of site, as shown in Figure 7-20.**

Figure 7-20:
The electronics concealed in the puppet.

Trying It Out

At this point, the puppet is ready to go. Press a hand or the little guy's nose, and listen to what he has to say.

If something doesn't work, here are the obvious things to check out:

- ✔ Check that all the batteries are fresh, tightly inserted in the battery pack, and all face the right direction.
- ✔ Check to see that no wires or components have come loose.

Taking It Further

Adding sound to objects has endless potential, as manufacturers of stuffed toys and dolls have discovered. If you like this project, here are some ways to vary it or take it further:

- ✔ **You can use a different puppet or stuffed animal or plastic toy in which to put the project.**
- ✔ **Create a music box that plays when you press a switch.**
- ✔ **Add more switches for people to push.**

  You can use up to eight of the event pins for sound.

- ✔ **Add light to sound.**

  For example, you might want to put a light on Merlin's head that lights up every time he speaks. See Chapter 5 for some ideas about working with LEDs.

If you want to program the SpeakJet only once, you don't need to keep the MAX232 with your project. Instead, you can move the SpeakJet to a smaller breadboard. That way, you can fit the board into a smaller puppet or toy.

Chapter 8

Surfing the Airwaves

· ·

In This Chapter

▶ Taking a close look at the radio's circuit

▶ Checking off the parts list

▶ Putting the radio circuit on a breadboard

▶ Assembling the radio box

▶ Tuning in!

· ·

*F*or years, people attracted to electronics have been drawn to radios. In olden days, radios had big tubes and were the size of a bookcase; today, you can create a small radio yourself by using common electronics components.

The radio in this project involves a simple circuit, speaker, and a coil and variable capacitor combo that you use to tune into your favorite station.

By working on this project, you discover how to work with radio frequency and amplification to grab radio signals out of the air.

The Big Picture: Project Overview

After you complete this project, you'll have an AM radio. We know, our favorite stations are on FM, too, but AM *(amplitude modulation)* was the first method of modulating frequencies used in radio. Building an AM radio is a bit easier to do, so that's where we start you off.

If you're dying to discover more about creating an FM *(frequency modulation)* radio, check out *Basic Radio: Understanding the Key Building Blocks,* by Joel Hallas, published by the American Radio Relay League (www.arrl.org).

You can see the finished radio in Figure 8-1.

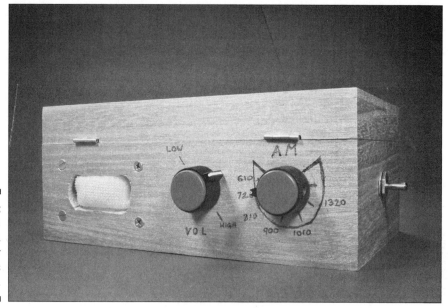

Figure 8-1:
Painted a
shiny silver,
we think our
radio is
cool.

Here are the types of activities you'll be doing to create your own radio. You will

1. Put together an electronic circuit containing a coil and variable capacitor that you use to tune to your favorite stations, an IC that separates the voice signal from the radio frequency carrier and an IC that amplifies the voice signal enough to power a speaker.

2. Install the circuit, speaker, on/off switch, volume control, and tuner in a handy box.

Scoping Out the Schematic

Imagine all that music and talk floating around out there in the air. Time to get to work on a circuit that lets you capture those sounds and pump them through a speaker. There is only one breadboard to put together for this project. You can see the schematic for the board in Figure 8-2.

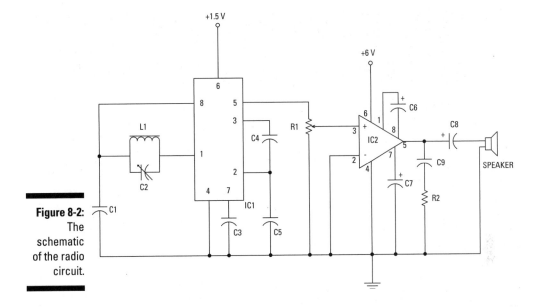

Figure 8-2:
The
schematic
of the radio
circuit.

The following is a list of the schematic elements for your radio:

- **L1** is a coil, or inductor. You make this by wrapping wire around a ferrite rod. This coil is both the antenna and half the tuning circuit needed to tune the radio to a particular station. The value of the inductor, given in Henrys, is determined by the number of coils of wire you wrap around the rod.

- **C2** is a variable capacitor that forms a tuning circuit along with L1. When you have a capacitor and inductor in parallel, the value of the capacitor and inductor determine the *resonance frequency,* which is the frequency of the radio station you tune in on your radio. As you change the value of C2 by turning the knob connected to the capacitor, you change the resonance frequency, therefore tuning in to a different radio station.

- **IC1** is a ZN416E integrated circuit that is designed to separate the voice signal from the radio frequency carrier and amplify the signal to a level sufficient to power headphones.

- **C1** is a capacitor that allows nonresonant radio frequency signals to conduct to ground.

- **C3, C4, and C5** are suggested by the manufacturer of IC1; they shunt high-frequency signals to ground, preventing them from causing noise in certain parts of the circuit inside of IC1.

- **IC2,** an audio amplifier named LM386N-1, takes the audio frequency electrical signal generated by IC1 and amplifies it to provide sufficient power to drive the speaker.

Picking up the right signal

Ferrite, used for the coil core in this project, is composed of crystalline iron oxide and is typically used at radio frequencies. For coils used at low frequencies, such as in power transformers, you can use iron as the core; however, at high frequencies, the high conductivity of iron allows eddy currents to be created in the core. These eddy currents can decrease the level of signal induced to the coil, thus weakening the signal. Because these eddy currents don't occur in the lower conductivity ferrite core, you get a much more efficient coil at radio frequencies.

When a radio signal comes across a ferrite rod, the radio signal creates a magnetic field in the rod. When you wrap a coil of wire around the ferrite rod, this magnetic field — which is changing direction at high frequency (the frequency of the radio signal) — induces (hence the name *inductor*) an electric current in the wire coil. A magnetic or electric field can cause electrons to flow in a wire; if the field is changing direction frequently, like one created by a radio signal does, the direction of the electrons changes with the magnetic field. *Voilà!* You now have a radio frequency signal traveling through your wire coil.

Because the inductor and capacitor are connected, this radio frequency signal also travels through the capacitor. When you have an inductor and capacitor in parallel, like in our AM radio circuit, the values of the inductor and capacitor determine a resonance frequency. A signal at the resonance frequency is then blocked from moving through the inductor/ capacitor circuit. This resonance frequency is the frequency of the radio station that you tune into when you change the value of the variable capacitor by turning the knob on your radio.

Signals from other radio stations that aren't at the resonance frequency flow through the inductor/capacitor circuit and through C1 to ground; basically, these signals are thrown away.

Only the signal at the resonance frequency doesn't go to ground through C1; rather, that signal is available to your circuit to be processed back into the music your radio station sends out.

- **R1** is a potentiometer that controls the sound volume.

- **C6** sets the voltage gain of IC2 to 200. (Therefore, the voltage out will be 200 times the voltage in.)

- **C7** improves the stability of the LM386 amplifier to prevent problems such as oscillation, which can turn your signal into an unintelligible garble of sound.

- **C8** removes any DC offset from the output of the LM386 amplifier.

- **C9** acts as a current bank for the output. This capacitor drains when sudden surges of current occur and refills with electrons when the demand for current is low.

Building Alert: Construction Issues

The ferrite rod that you wrap wires around to make a coil is brittle. If you drop it or knock it against something, it could easily break, so be careful!

If you place your radio in a wooden box (as we did), check the thickness of the wall of the box against the length of the threads on the potentiometer. Because the wall on our box was ¼" thick (the same thickness as the threads on the potentiometer), we just put some glue between the body of the potentiometer and the wall of the box (making sure not to get glue on the rotating shaft). You might want to use a chisel to reduce the thickness of the wall so that the rotating shaft sticks out a little farther; that way, the knob doesn't end up too close to the face of the box.

The variable capacitor has very small holes in its metal body that you can use to mount it on small pins. Again, we found it easiest to glue this to the wall of the box after *chamfering* (cutting a furrow in) the edge of the hole to fit the metal knuckle around the shaft. In this way, the wall of the capacitor fits flush against the wall of the box.

Perusing the Parts List

Get ready to turn on your computer or head to the electronics supply store to shop for the parts for your radio project (several of which are shown in Figures 8-3 and 8-4). Here's what you'll need:

- ✔ **10 kohm potentiometer (R1)**

- ✔ **10 ohm resistor (R2)**

- ✔ **Two 0.01 microfarad ceramic capacitors (C1, C3)**

- ✔ **One 0.1 microfarad ceramic capacitor (C5)**

- ✔ **Two 0.047 microfarad ceramic capacitor (C4, C9)**

- ✔ **10 microfarad electrolytic capacitors (C7, C6)**

- ✔ **100 microfarad electrolytic capacitor (C8)**

- ✔ **LM386N-1 amplifier IC2**

    You'll find many versions of the LM386 amplifier. We chose the LM386N-1 because it works with the supply voltage of 6 volts used by this circuit.

- ✔ **Battery pack for 1 AA battery**

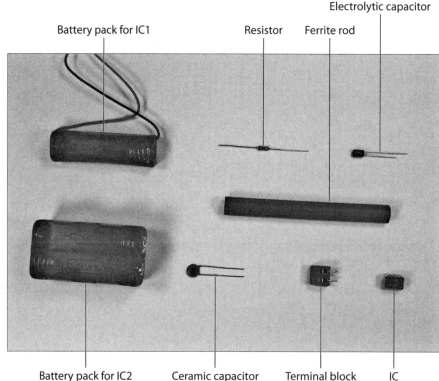

Battery pack for IC1   Resistor   Ferrite rod   Electrolytic capacitor

Figure 8-3:
Key
components.

Battery pack for IC2   Ceramic capacitor   Terminal block   IC

This supplies 1.5 volts to IC1 because the maximum supply voltage for the ZN416 is 1.6 volts.

✔ **Battery pack for 4 AA batteries to supply 6 volts to IC2**

✔ **ZN416E AM radio receiver IC1**

✔ **14–365 picofarad variable air capacitor (C2)**

You could also use a variable capacitor that goes up to 500 picofarad.

✔ **⅜"-diameter × 3½" ferrite rod**

You can also use a ferrite rod with a ½" diameter and longer lengths of either ⅜" or ½" diameter as long as they fit in the box that you're using for the radio. Longer ferrite rods should give you higher sensitivity to weak radio signals than shorter ferrite rods.

We found the preceding three radio-specific items above at Ocean State Electronics (www.oselectronics.com). We noticed that it was less expensive to buy this vendor's part #LA-540 — a ferrite rod already wrapped with a wire coil — slip off the coil, and then wrap wire to the number of turns we needed on that rod. Buying a ferrite rod by itself was about three times as expensive!

On/off switch   Speaker

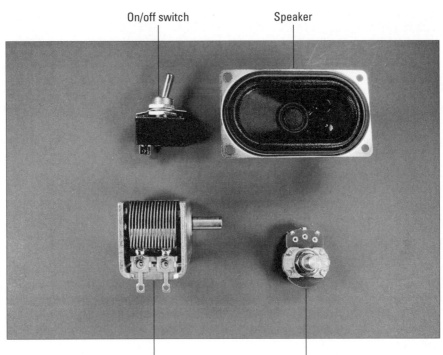

Figure 8-4:
More key
components.

Variable capacitor   Potentiometer

✒ **SPST toggle switch, used as the on/off switch**

✒ **830-pin breadboard**

✒ **Seven 2-pin terminal blocks**

✒ **Two knobs (for the potentiometer and variable capacitor)**

✒ **8 ohm, 1 watt speaker**

✒ **Enclosure**

We used a wooden box that we found at a national craft supply store (Michael's). You can use plastic or wood but *do not use metal* because it will block the radio signal from the ferrite antenna.

✒ **Four ½" long 6-32 flathead screws**

✒ **Four 6-32 nuts**

✒ **An assortment of different lengths of prestripped, short 22 AWG wire**

✒ **A few feet of 26 gauge enamel-coated wire (used to make the coil)**

Taking Things Step by Step

If you have one, now's the time to put on your official Marconi Was Here T-shirt so you can set out to build your AM radio in style. Building a radio involves creating the circuit; drilling various holes in a box; placing the circuit, speaker, and knobs in the box; and tuning to find available frequencies.

Building a radio circuit

The first step in building a radio is to tackle the circuit that forms its electronic brains. Here are the steps involved:

1. **Place ZN416E (IC1), LM386N-1 (IC2), and seven terminal blocks on the breadboard, as shown in Figure 8-5.**

    The seven terminal blocks shown in this figure will be used to connect two wires each to various components in the circuit. The wires from these terminal blocks go to the battery pack for IC1, the battery pack for IC2, the on/off switch, the coil, the variable capacitor, the speaker, and the potentiometer.

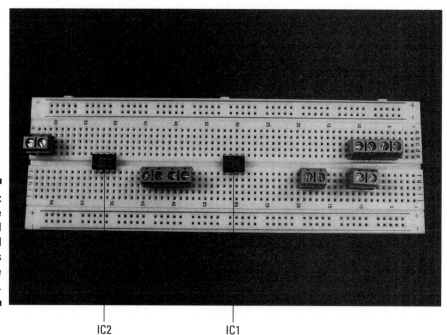

Figure 8-5: Place the ICs and terminal blocks on the breadboard.

IC2   IC1

2. **Insert wires to connect the ICs and the terminal blocks to the ground bus and insert a wire between the two ground buses to connect them to each other, as shown in Figure 8-6.**

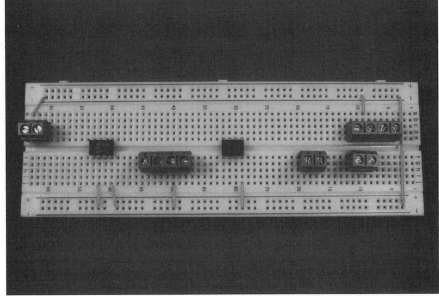

Figure 8-6:
Six shorter wires connect components to ground bus; the long wire on the right connects the two ground buses.

3. **Insert wires to connect IC2 and the terminal block for IC2's battery pack to the +V bus, and a wire between the two +V buses to connect them to each other, as shown in Figure 8-7.**

   Do not connect IC1 to the +V bus. If you do, you will probably fry the poor IC because it isn't designed to take 6 volts.

4. **Insert wires to connect the ICs, terminal blocks for the coil (L1), variable capacitor (C2) and terminal blocks for the potentiometer (R1), and terminal block for the speaker and discrete components, as shown in Figure 8-8.**

5. **Insert two 0.047 microfarad capacitors (C4 and C9), two 10 microfarad capacitors (C6 and C7), one 100 microfarad capacitor (C8), two 0.001 microfarad capacitors (C1 and C3), one 0.1 microfarad capacitor (C5), and one 10 ohm resistor (R2) on the breadboard, as shown in Figure 8-9.**

   When inserting electrolytic capacitors, be sure to check the schematic to see where to insert the longer, positive (+) lead.

Pin 6 of IC2 to +V   Battery terminal block to +V

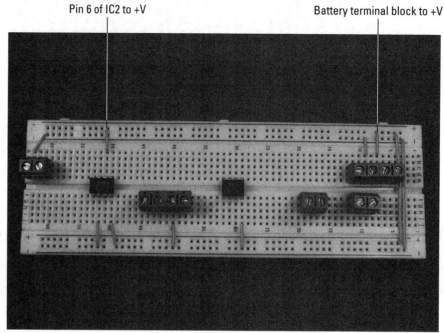

Figure 8-7:
Connect
components
to the +V
bus.

We discuss in Chapter 4 how you can determine how short to clip the leads of many of these components to make them fit neatly on the breadboard. Not to sound like your mother, but make sure you wear your safety glasses when clipping leads!

Making a box into a radio

After you build the circuit, it needs a home, so the next step in the process is to work with the box you purchased for your radio enclosure.

Follow these steps to get the box radio ready:

1. **Drill holes in the box where you will mount the variable capacitor, potentiometer, and on/off switch.**

   We put both the on/off switch on one side of the box and the speaker, potentiometer and variable capacitor on another side, but the placement is really up to you. Figure 8-10 shows where we placed these components.

Open region to speaker TB

Pin 5 of IC2 to open region

Pin 5 of IC1 to R1 TB

L1/C2 TB to Pin 8 of IC1

Battery pack TB to Pin 6 of IC1

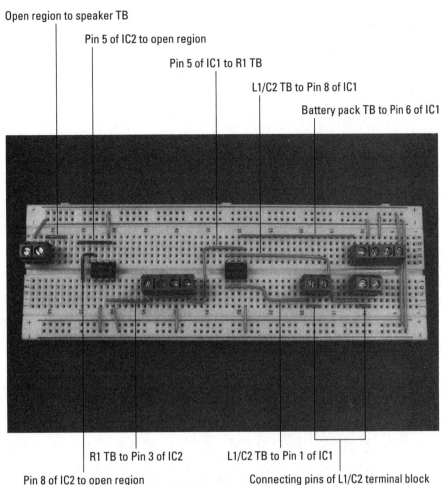

Figure 8-8:
Hook up the
ICs, terminal
blocks (TB),
and discrete
components.

R1 TB to Pin 3 of IC2

L1/C2 TB to Pin 1 of IC1

Pin 8 of IC2 to open region

Connecting pins of L1/C2 terminal block

2. **Place the speaker against the face of the box where you want to mount it and mark positions for the four screws you'll use to secure the speaker.**

3. **Drill holes to allow clearance for 6-32 screws.**

   We used a ⁹⁄₆₄" drill bit.

4. **Draw an outline about ¼" smaller than the speaker shape for the opening to let sound from the speaker out.**

   Use a coping saw to cut the opening.

C8 from IC2 Pin 5 to speaker TB

C7 from Pin 7 of IC2 to ground

C9 from Pin 5 of IC2 to open region

R2 from C9 to ground

C1 from IC1 Pin 8 to ground

C3 from IC1 Pin 7 to ground

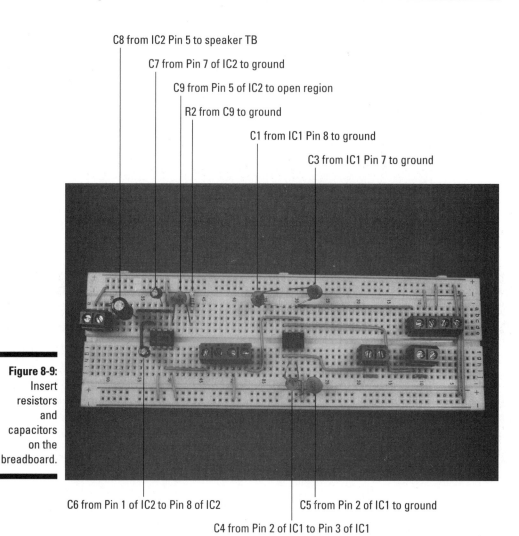

Figure 8-9:
Insert
resistors
and
capacitors
on the
breadboard.

C6 from Pin 1 of IC2 to Pin 8 of IC2

C5 from Pin 2 of IC1 to ground

C4 from Pin 2 of IC1 to Pin 3 of IC1

See Chapter 4 for more information about choosing drill bit sizes for particular components and other pieces of wisdom on how to customize a box for your projects. Make sure you use safety glasses when drilling and clamp the box to your worktable!

5. **Slip the shaft of the on/off switch through the drilled hole and secure with the nut provided.**

6. **Slip the shaft of the potentiometer through the drilled hole and secure with the nut provided.**

Potentiometer

Variable capacitor

Speaker

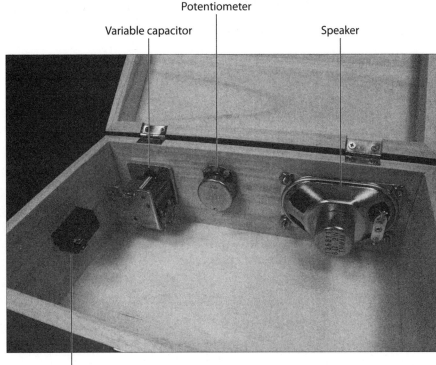

Figure 8-10:
Box with on/
off switch,
potenti-
ometer,
variable
capacitor,
and speaker
in place.

On/off switch

7. **Slip the knob on the potentiometer shaft and secure with the set screw provided.**

   The tread on potentiometers is about ¼" long, so if the wall of your wooden box is ¼" thick, you won't be able to use the nut to secure the potentiometer. Instead, glue the face of the potentiometer to the box, making sure that you don't get any glue on the rotating shaft of the potentiometer.

8. **Slip the shaft of the variable capacitor through the drilled hole and glue the metal body of the capacitor to the wooden box.**

   Make sure you don't get any glue on the shaft or other moving parts of the capacitor.

9. **Slip the knob on the variable capacitor shaft and secure with the set screw provided.**

10. **Secure the speaker with four 6-32 flathead screws and four 6-32 nuts.**

11. **Solder the black wire from each of the battery packs to one lug of the on/off switch and solder an 8" black wire to the remaining lug of the on/off switch, as shown in Figure 8-11.**

12. **Solder one 8" wire to a variable capacitor lug and solder one 8" wire to the metal body of the variable capacitor, as shown Figure 8-11.**

    The lugs are electrically connected to the stationary plates of the capacitor, and the metal body is electrically connected to the rotating plates.

Black wire to on/off switch   Wire to capacitor body

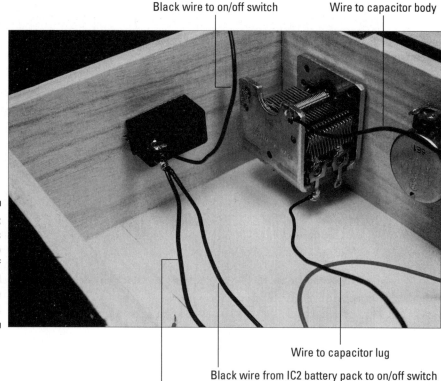

Figure 8-11:
Wires soldered to the on/off switch and variable capacitor.

Wire to capacitor lug

Black wire from IC2 battery pack to on/off switch

Black wire from IC1 battery pack to on/off switch

13. **Solder 8" wires to each of the three potentiometer lugs, as shown Figure 8-12.**

14. **Solder 8" wires to each of the two speaker lugs, as shown Figure 8-12.**

Wires to potentiometer

Figure 8-12:
Wires
soldered to
potenti-
ometer and
speaker.

Wires to speaker

Be sure to heed all the safety precautions about soldering that we give you in Chapter 2. For example, don't leave your soldering iron on and unattended. And just in case a bit of solder has an air pocket and pops, wear your safety glasses!

Coaxing the coil

Take your ferrite rod, 26 gauge enamel-coated wire, some electrical tape, and glue and get ready to make the coil by following these steps:

1. **Wrap electrical tape around the ferrite rod.**

   This protects the enamel coating on the wire from the ridges that run along the rod.

2. **Cut two small pieces of electrical tape, about ¼" wide, and place them in a handy spot.**

3. **Leaving about 8 inches of wire loose, start winding the enamel-coated wire around the rod, starting about 1¼" from one end.**

4. **When you have about 20 turns of wire on the rod, secure that end with one of the pieces of electrical tape so that it doesn't loosen while you are finishing the coil.**

5. **Wind a total of 50 turns of wire around the rod, leaving another 8-inch length of wire loose at the other end.**

6. **Secure the wire at that end with the other piece of electrical tape.**

7. **At each end of the coil, use some glue to hold down the wires.**

   We had you cut the electrical tape ¼" wide so there is enough exposed wire to glue. When the glue is dry, your coil is finished. The finished coil is shown in Figure 8-13.

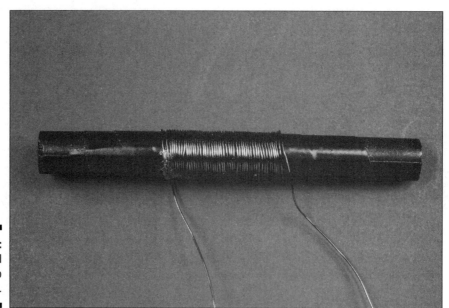

Figure 8-13:
Your coil, all ready to use.

Putting it all together

Now it's time to assemble your radio and see whether you can pick up a signal. Follow these steps to complete the project:

1. **Attach Velcro to the breadboard and the box and then secure the breadboard in the box.**

2. **Attach Velcro to the battery packs and the box and then secure the battery packs in the box.**

3. **Insert the wires from the speaker, potentiometer, variable capacitor, coil, battery packs, and the on/off switch to the terminal blocks on the breadboard, as shown in Figure 8-14.**

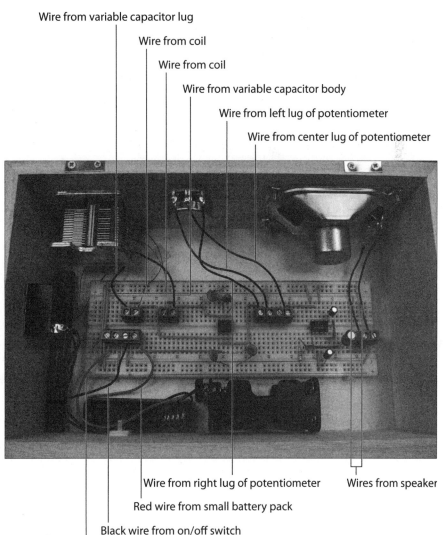

Wire from variable capacitor lug

Wire from coil

Wire from coil

Wire from variable capacitor body

Wire from left lug of potentiometer

Wire from center lug of potentiometer

Figure 8-14:
Connect the speaker, variable capacitor, on/off switch, and battery packs to the breadboard.

Wire from right lug of potentiometer

Wires from speaker

Red wire from small battery pack

Black wire from on/off switch

Red wire from large battery pack

4. **As you insert the wires, cut each of them to the length needed to reach the assigned terminal block and strip the insulation from the end of the wire.**

   Keep the wires from the potentiometer as far away as possible from the wires from the speaker or the wires from the variable capacitor/coil. You've no doubt experienced how having a microphone too close to a speaker can produce an awful screech; the same screech can occur if these wires get too close together.

5. **Secure the wires with wire clips where needed.**

   The way parts are laid out in this box and the distance between some of the components (such as the speaker, potentiometer, and variable capacitor to the terminal blocks) is short enough that you won't need wire clips.

6. **Close the lid on the box and admire your finished radio, as shown in Figure 8-15.**

Figure 8-15:
The finished radio.

Trying It Out

At this point, the radio is ready to go. Insert batteries and turn it on. Adjust the tuner (the knob on the variable capacitor) and the volume (the knob on the potentiometer) to find listen to your favorite radio stations.

When you tune into a station that you want to find again, mark the position of the tuning knob on the face of the radio.

If you're not getting the signal you thought you would, here are the obvious things to check out:

✔ **Check that all the batteries are fresh and tightly inserted in the battery pack, all facing the right direction.**

✔ **Check to see that no wires or components have come loose.**

✔ **Rotate the radio.**

When one end of the antenna points at the radio station's transmitting antenna, the signal from that radio station will be stronger.

Taking It Further

Many people get bitten by the radio bug and want to get into ham radio in a big way. If you're one of these, here are some variations on this project to keep you going:

✔ You can build an FM radio to get those high-frequency stations. The TEA5710N IC can be used to build a radio that can receive both AM and FM signals.

✔ Build a radio that will receive signals from ham radio operators.

Go to QRP/SWL HomeBuilder (www.qrp.pops.net/default.htm) for some ideas for ham radio projects, or visit www.arrl.org.

✔ Mount an external antenna connected to the ferrite rod to boost the power of the signal if you live in the boondocks where signals are hard to come by.

Part III
Let There Be Light

In this part . . .

Thomas Edison discovered what fun light can be in electronics projects when he invented the light bulb. In the projects in this part, you use light in several forms. For example, you work with tiny LED lights that dance across a display of dolphins; use infrared light to signal a go cart when to go and when to stop; and use an infrared beam to detect motion to set off lights and sound in a snazzy Halloween display.

Chapter 9

Scary Pumpkins

· ·

· ·

*A*round the end of October, many of us carve faces in pumpkins and then put candles in those pumpkins to create an eerie effect. Why not electrify pumpkins to do the same thing — and add a spooky sound effect or message to the mix?

In this chapter, we use two plastic pumpkins and activate sound and light by using an infrared beam. When those trick-or-treaters come up to your doorway, won't they be surprised?

Of course, if it's February and plastic pumpkins are scarce, use these same techniques with some other plastic container shape to create talking dinosaurs, heart-shaped candy boxes, or whatever!

The Big Picture: Project Overview

When you complete this project, you'll have two pumpkins:

- ✔ One that transmits an infrared beam
- ✔ A second one that lights up and plays back a recorded sound or message when something or somebody interrupts the infrared beam by walking between the two pumpkins

You can see the finished pumpkin in Figure 9-1.

Figure 9-1:
The final
product: a
talking
pumpkin.

Here's the big picture of this pumpkin project:

1. Put together two electronic circuits and fit them into plastic pumpkins with switches, a microphone, and a speaker.

2. Use a microphone to record a sound or message.

   We like, "Welcome to Sleepy Hollow. We hope you have a good time," followed by a spooky laugh.

3. One pumpkin transmits an infrared beam to the other. When someone walks between the pumpkins, the recorded message is triggered, along with a flickering red light.

4. To reduce the chance of this IR noise interfering with your gadget, use an IR detector tuned to detect infrared that turns on and off at 38 kHz and ignores infrared not switched at that frequency. The transmitter circuit then sends out infrared that is switched on and off at 38 kHz, and the noise problem is solved.

One complication you'll deal with along the way is lots of infrared noise floating around. IR is given off by heaters, people, pets, and pretty much any living creature or equipment that gives off heat.

Should you be lucky enough to have an oscilloscope perched on your workbench, you can see that a 38 kHz square wave looks something like the one shown in Figure 9-2. But don't worry: You don't have to have an oscilloscope to tune your IR transmitter. We tell you how to tune it in the upcoming section, "Trying It Out."

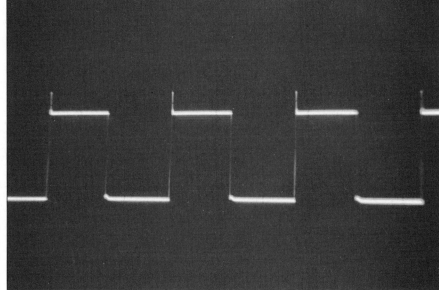

Figure 9-2:
A square wave on an oscilloscope screen looks like this.

Scoping out the schematic

You put together two breadboards for this project: one that transmits and one that receives and sounds off.

First, you can see the schematic for the board that goes in what we call the *silent pumpkin* (the one with the transmitter in it) in Figure 9-3.

Here's the nitty-gritty of the schematic elements for the silent pumpkin:

✔ The **IR LED (LED2)** is one of the key components of this circuit; the purpose of the rest of the circuit is to send an electrical current, which turns on and off at a frequency of 38 kHz, through this LED. This current causes the LED to transmit IR light that turns on and off at a frequency of 38 kHz (38,000 times a second: so fast you can't even see a flicker).

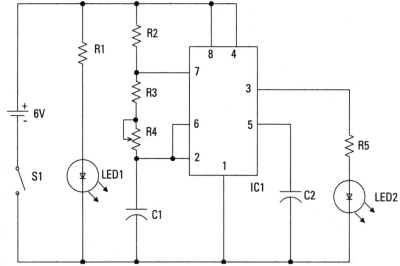

Figure 9-3:
The silent
pumpkin
circuit
schematic.

✔ **IC1** is the other key component of this circuit. This is an LM555 timer chip that you use to generate a square wave at its output on Pin 3.

✔ **R2, R3, R4, and C1** are three resistors and a capacitor, respectively, that form the RC circuit that determines the frequency of the square wave generated by the LM555 timer chip.

✔ **S1** is an SPST (single-pole, single-throw; see Chapter 4) toggle switch connected between the negative pole of the battery pack and the bread-board ground bus. When this switch is open, no current can flow, and so the circuit turns off. When this switch is closed, the circuit turns on.

✔ **LED2** provides a light (we use a yellow light) to simulate a candle's glow in the pumpkin. This LED is on whenever S1 is closed.

✔ **R1** is a resistor that limits the current running through LED1 to approximately 20 milliamps (mA).

✔ **R5** is a resistor that limits the current running through LED2 to approximately 30 milliamps (mA).

✔ **C2** is a capacitor that reduces the occurrence of noise on Pin 5, which could cause false triggering of the IC. This might occur if Pin 5 were left unconnected.

Now it's time to run down the elements of the receiver schematic that goes into the talking pumpkin. Take a look at the schematic in Figure 9-4.

Time, time, timers

When you connect a 555 timer IC to resistors and capacitors in the arrangement shown in the schematic, the timer IC generates a digital waveform from its output. The frequency of the wave-form is determined by how fast the capacitor fills and drains. You calculate how fast the capaci-tor fills to two-thirds of its capacity or drains to one-third of its capacity by using the *RC time constant equation.* (This involves math, so it's not for the faint of heart.)

The RC time constant for filling the capacitor is

$T1 = (R2 + R3 + R4) \times C$

The RC time constant for draining the capacitor is

$T2 = (R3 + R4) \times C$

In this circuit, R2, R3, and R4 determine how fast the capacitor charges and discharges. The extent to which the capacitor is filled determines the voltage on Pins 2 and 6 and the voltage applied to the circuit inside the IC. When the voltage reaches two-thirds of +V, the circuitry connected to Pin 6 turns on and causes the output to change from +V to 0 (zero) volts. It also causes the charge on the capacitor to drain through Pin 7 to ground. As the capacitor drains, the voltage to Pins 2 and 6 drops. When the voltage gets to one-third of the +V, the circuitry connected to Pin 2 turns on and causes the output of the IC to shift from 0 (zero) to +V and disconnects Pin 7 from ground, which allows the capacitor to charge back up to two-thirds of +V. At this point, the cycle starts again.

✔ The **IR detector** is the key component of this circuit. It contains a photo-diode that detects infrared light and an integrated circuit that produces either +V or 0 volts on its output pin. Exactly what volts the IR detector produces depends on whether it detects a 38 kHz infrared signal (result-ing in 0 volts output) or not (resulting in +V output).

✔ **IC1** is the other key component of this circuit. This is a chip that you can use to record a sound or voice message and play it back. We con-nect the output of the IR detector to Pin 23 of IC1. Voltage on Pin 23 starts a playback when the voltage changes from +V to 0 volts. Here's how this works: When a person walks between the pumpkins, the volt-age from the IR detector changes from 0 volts to +V. When the person leaves the beam field, it drops back down to 0 volts. The jump back to 0 volts is the point when your recording starts to play.

✔ The **speaker** is connected to Pins 14 and 15 of IC1. The speaker plays messages that you recorded on IC1.

✔ You connect **LED1** between Pin 14 of IC1 and ground. When your mes-sage plays, this LED generates a flickering light. (We used a red LED to get a red light.)

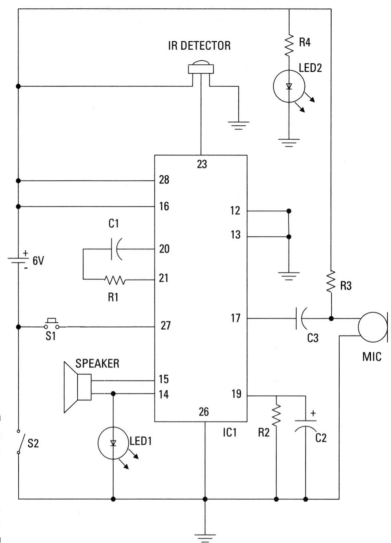

Figure 9-4:
The
schematic
of the
talking
pumpkin
circuit.

For a brighter light, try using an LED with a clear glass shell instead of a translucent shell.

✓ **LED2** provides a steady light.

We chose yellow, as if you had a candle in the pumpkin.

✓ **R4** is a resistor that limits the current running through LED2 to approximately 20 milliamps.

- ✔ **S1** is a normally open (NO) pushbutton switch that when depressed, connects Pin 27 of IC1 to ground. This is how you record sounds to IC1 through the microphone. Recording stops when you release the S1 pushbutton.

- ✔ **R3** is a resistor that connects the microphone to +V, supplying the 4.5 volts that the microphone needs to function.

- ✔ **C3** is a capacitor that removes the DC voltage from the AC signal that's flowing from the microphone to Pin 17 of IC1.

- ✔ **S2** is the on/off switch between the negative terminal of the battery pack and the ground bus of the circuit board.

- ✔ **R1** and **C1** filter out that pesky electrical noise.

- ✔ **R2** and **C2** connect the automatic gain control circuit inside IC1 to ground. The values of R2 and C2 determine how fast the automatic gain control responds to changes in volume when you're recording a message.

Building alert: Construction issues

Plastic pumpkins present unique building challenges, as everybody (who bothers to think about it) knows. The curved shape of the pumpkin makes it difficult to stick your hands inside to attach wires to terminal blocks. Therefore, we attached the wires from most components — like switches and speakers — to terminal blocks on the breadboard before placing them in the pumpkin or attaching the components to the side of the pumpkin.

Wrap each lead of the IR detector and microphone in electrical tape after they have cooled from soldering to make sure that the bare soldered leads don't touch each other. Leads co-mingling in this fashion can cause minor electrical disasters.

We used a utility knife and small wire cutters to cut the holes in the plastic pumpkins. Make sure that the plastic is not too brittle, and be sure to wear safety goggles to protect your eyes against flying pumpkin pieces and leather work gloves to protect your hands.

We placed some foam in the bottom of the pumpkins on which to rest the breadboards. The type of foam used for flower arrangements is easy to cut, but it's so fine that particles go everywhere. The foam that you find in packing boxes will also shred, but the pieces are bigger and a little easier to clean up. In either case, plan to spend some time cleaning up your cutting surface afterward and don't cut on your electronics bench — you don't want those foam pieces to get into the electronics bits and pieces.

Perusing the parts list

In addition to buying the two plastic pumpkins, you have to find the electronic parts to build the circuits and assemble everything. Here are the parts lists for the silent pumpkin and the talking pumpkin.

Silent pumpkin parts list

It's time to go shopping for the bits and pieces used to build the silent pumpkin. This involves the following parts (several of which are shown in Figure 9-5):

- ✔ **330 ohm resistor (R1, R2)**
- ✔ **10 kohm resistor (R3)**
- ✔ **10 kohm potentiometer (R4)**
- ✔ **150 ohm resistor (R5)**
- ✔ **0.001 microfarad ceramic capacitor (C1)**
- ✔ **0.1 microfarad ceramic capacitor (C2)**
- ✔ **Yellow size T-1 ¾ LED (LED1)**
- ✔ **IR LED, TSAL7200 (LED2)**

    Various other IR LEDs will work. The one we used is advertised to work at longer distances than standard IR LEDs. We used this one along with the IR detector specified in the next section because it was easier to get the gadget to work at a distance up to about 15 feet than with other combinations we tried. See www.rentron.com for a nice listing of similar IR LEDs.

- ✔ **LM555 (IC1)**
- ✔ **SPST toggle switch (S1), used as the on/off switch**
- ✔ **400-contact breadboard**
- ✔ **LED socket, sized for T-1 ¾ LEDs**
- ✔ **Four AA battery pack with snap connector**
- ✔ **Three 2-pin terminal blocks**
- ✔ **Knob (for potentiometer)**
- ✔ **Foam (Styrofoam or similar), about 2½" thick, 12" wide, 12" long**
- ✔ An assortment of different lengths of prestripped short 22 gauge (AWG) wire

On/off switch   LM555 (IC1)   Battery pack   Resistor

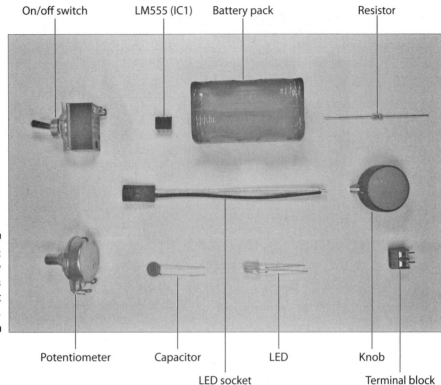

Figure 9-5:
Key
components
for the silent
pumpkin.

Potentiometer   Capacitor   LED   Knob

LED socket   Terminal block

Talking pumpkin parts list

Here's your shopping list for building your talking pumpkin. This project involves the following parts, several of which are shown in Figure 9-6:

- ✔ **Panasonic PNA4602 IR detector**

- ✔ **16 ohm, 0.2 watt speaker**

- ✔ **Electret microphone part #EM-99**

  We found this one at Jameco (www.jameco.com). You can use other electret microphones; see Chapter 3 for the criteria to help you choose one. If you use another model of microphone, you might have to adjust R3 to set the supply voltage to the correct level.

- ✔ **Winbond Electronics ISD1110 voice record/playback chip (IC1)**

- ✔ **0.01 microfarad capacitor (C1)**

- ✔ **0.1 microfarad capacitors (C3)**

Voice chip   Record switch   Battery pack   Speaker   Ceramic capacitor
Terminal block   On/off switch

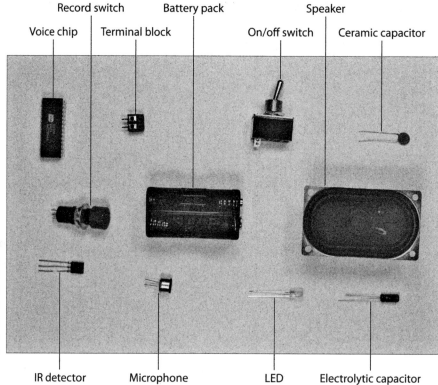

Figure 9-6:
Key
components
of the
talking
pumpkin.

IR detector   Microphone   LED   Electrolytic capacitor

- ✔ **4.7 microfarad electrolytic capacitor (C2)**
- ✔ **Four ⅝" 6-32 screws**
- ✔ **Four 6-32 nuts**
- ✔ **5.1 kohm resistor (R1)**
- ✔ **470 kohm resistor (R2)**
- ✔ **2.2 kohm resistor (R3)**
- ✔ **330 ohm resistor (R4)**
- ✔ **Yellow T-1 ¾ LED (LED1)**
- ✔ **Red T-1 ¾ LED (LED2)**
- ✔ **830-contact breadboard**
- ✔ **Normally open (NO) momentary contact pushbutton switch, also referred to as** *record switch* **(S1)**

✔ SPST toggle switch, also referred to as *on/off switch* (S2)

✔ Four AA battery pack with snap connector

✔ Six 2-pin terminal blocks

✔ An assortment of different lengths of prestripped short 22 AWG wire

Taking Things Step by Step

Scary things don't just happen by themselves (things that go bump in the night notwithstanding). Building a scary pumpkin project requires that you do two things:

1. Create the silent pumpkin.

2. Create the talking pumpkin.

Making a silent pumpkin

Tackle the silent partner of this pumpkin duo first. Here are the steps involved:

1. **Place the LM555 IC and three terminal blocks on a breadboard, as shown in Figure 9-7.**

Figure 9-7:
Place the LM555 IC and terminal blocks on the breadboard.

The three terminal blocks shown in this figure will be used to connect two wires each to various components in the circuit. The wires from these terminal blocks go to the battery pack, IR LED, and potentiometer, respectively.

2. **Insert wires to connect the IC and the terminal blocks to the ground bus (marked with a – sign on this breadboard) and insert a wire between the two ground buses to connect them to each other, as shown in Figure 9-8.**

   Three shorter wires connect components to the ground bus; the long wire on the right connects the two ground buses.

Figure 9-8:
Connect components to the ground bus.

3. **Insert wires to connect the IC and the terminal block for the battery to the +V bus, as shown in Figure 9-9.**

4. **Insert wires to connect the IC, terminal blocks, and discrete components, as shown in Figure 9-10.**

5. **Insert discrete components on the breadboard, as shown in Figure 9-11.**

   Note that the shorter of the LED leads is inserted in the ground bus.

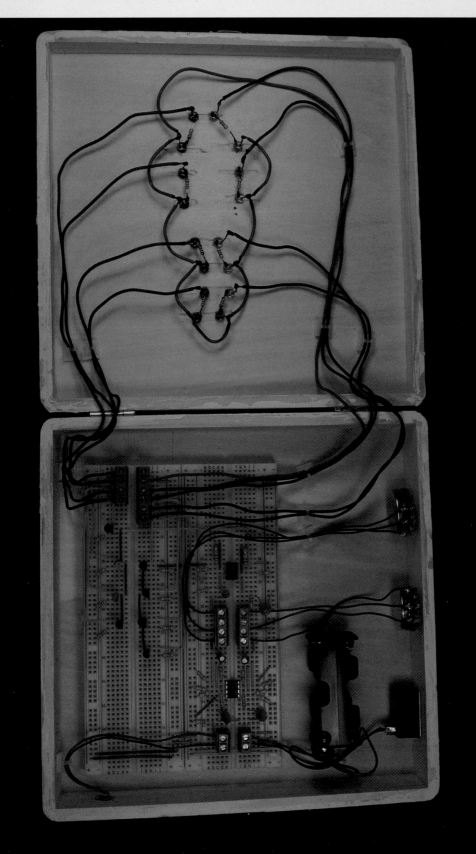

Here is the breadboard, wiring, and enclosure for the Dance to the Music project in Chapter 5. This project shows you how to separate out different sounds electronically with low-pass and high-pass filters.

This is the circuit for Sensitive Sam, whom you meet in Chapter 13. This maze of electronic gizmos can be controlled with a radio remote control in combination with sensors that allow Sam to follow a track using his own instincts.

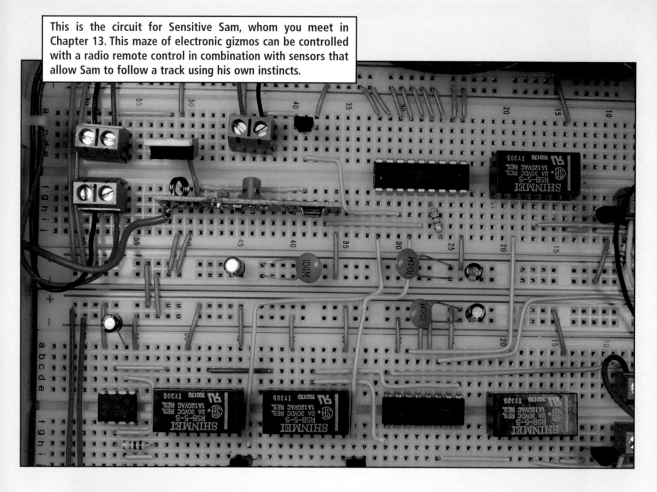

Here is Sam sensing his way around a track on the floor. This guy can move along a track on his own or at your command, speed up, and sound a horn.

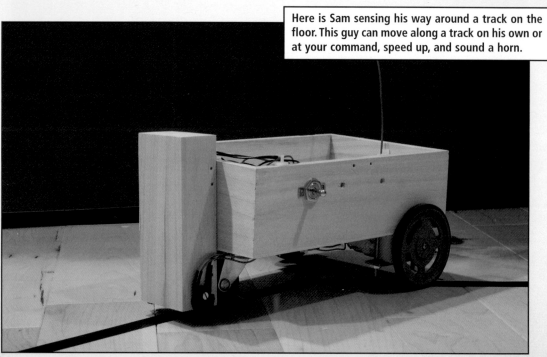

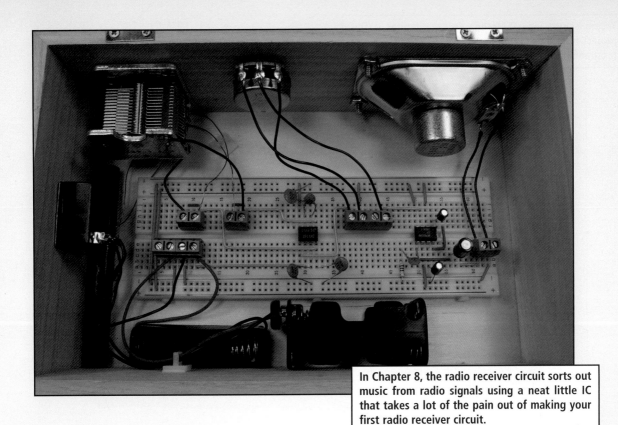

In Chapter 8, the radio receiver circuit sorts out music from radio signals using a neat little IC that takes a lot of the pain out of making your first radio receiver circuit.

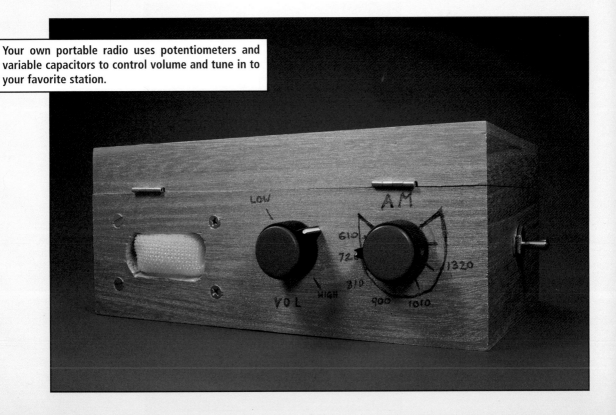

Your own portable radio uses potentiometers and variable capacitors to control volume and tune in to your favorite station.

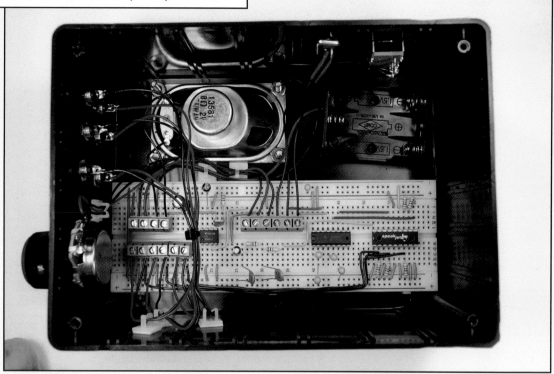

Taking a closer look at Merlin's insides (Chapter 7). This project uses a neat speech synthesizer IC that can be programmed with musical tones, words, and more.

This cute little guy has push buttons hidden here and there that you use to get him to talk.

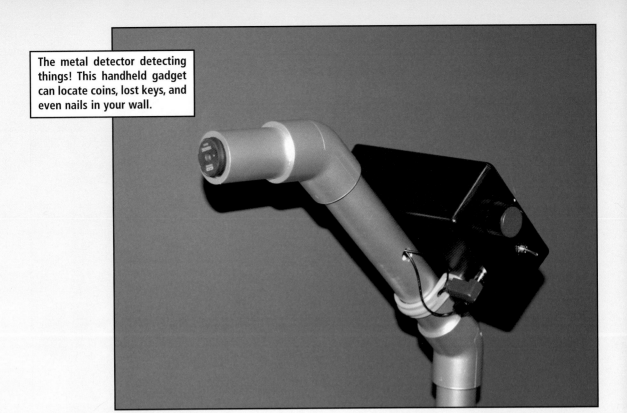

The metal detector detecting things! This handheld gadget can locate coins, lost keys, and even nails in your wall.

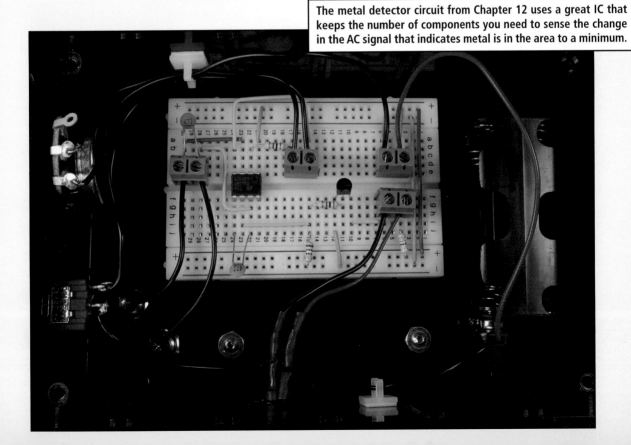

The metal detector circuit from Chapter 12 uses a great IC that keeps the number of components you need to sense the change in the AC signal that indicates metal is in the area to a minimum.

In Chapter 6, you build this circuit to drive the parabolic microphone that picks up sounds at a distance. Be careful with this circuit when placing the wires so that they don't get too close and cause feedback.

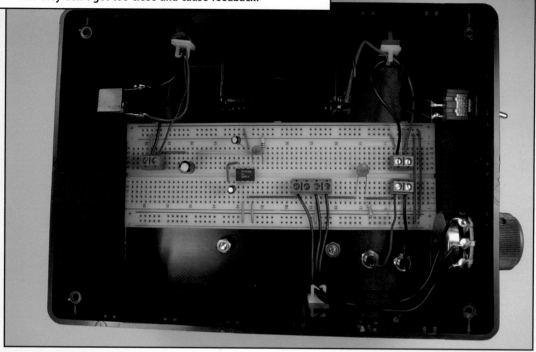

Chapter 11 is where you go to create this little bubble cart that you control with infrared light. This is just like controlling your TV, only you get to watch this guy dart around the room and go backward, which is way more entertaining than most TV shows.

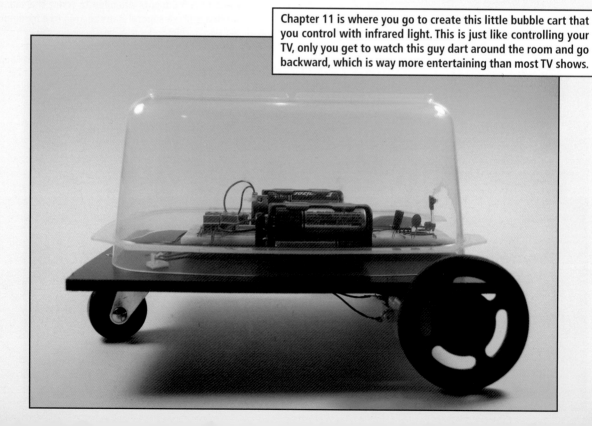

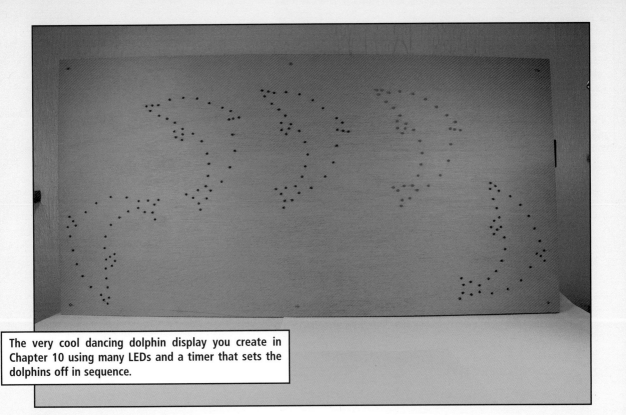

The very cool dancing dolphin display you create in Chapter 10 using many LEDs and a timer that sets the dolphins off in sequence.

The dancing dolphin circuit revealed. Each pair of wires goes to an array of LEDs that make up one dolphin. Wiring up the arrays is the time consuming part of this project!

A receiver circuit is housed in one plastic pumpkin, and another pumpkin houses the transmitter (see Chapter 9). When somebody walks between the two, a beam of infrared light is disrupted and a recorded message plays using a sound chip.

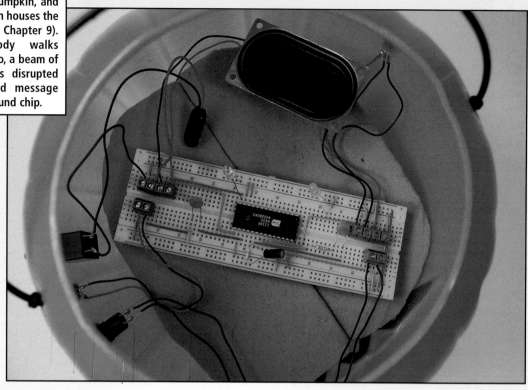

The Couch Petato circuit from Chapter 14 uses a vibration sensor to detect your pet when it jumps on a piece of furniture, and a sound chip to tell your cat to scat.

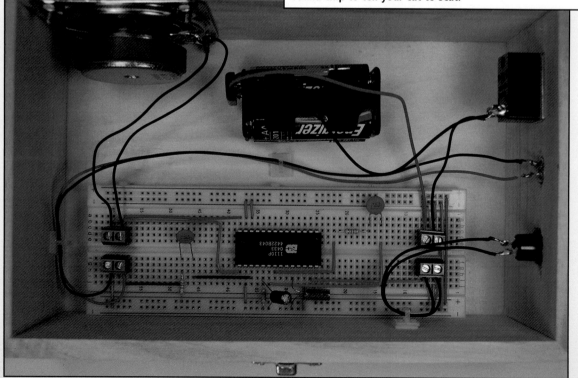

Pin 4 of IC1 to +V

Figure 9-9:
Connect
components
to the +V
bus.

Pin 8 of IC1 to +V   Battery terminal block to +V

6. **Solder the black wire from a battery snap to one lug of the on/off switch and solder a 5" black wire to the other lug of the on/off switch.**

7. **Solder two 5" wires to the potentiometer, as shown Figure 9-12.**

Be sure to heed all the safety precautions about soldering that we give you in Chapter 2. For example, don't leave your soldering iron on and unattended. And just in case a bit of solder has an air pocket and pops, wear your safety glasses!

8. **Cut the leads of the IR LED to ¼", keeping track of which is the long (+V) lead.**

Wear your safety glasses any time you cut wires!

9. **Insert the IR LED in the LED socket, making sure that the +V lead is lined up with the white wire from the socket.**

Figure 9-13 shows the LED inserted into the socket.

IC Pin 2 to IC Pin 6

LED terminal block to an open region

Figure 9-10:
Hook up the
IC, terminal
blocks, and
discrete
components.

IC Pin 6 to potentiometer terminal block

Potentiometer terminal block to an open region

10. **Attach wires from the LED, battery pack snap, on/off switch, and potentiometer to the terminal blocks, as shown in Figure 9-14.**

11. **Use a mini hacksaw or a utility knife to shape foam blocks so that they fit inside the bottom of the pumpkin, as shown in Figure 9-15.**

    We used the type of foam used to hold dried flower arrangements. You could also use packing foam.

12. **On the side of the plastic pumpkin that will face the talking pumpkin, cut a hole just large enough to allow the LED socket to fit.**

13. **On the side of the plastic pumpkin that will face away from visitors, cut holes just large enough to allow the shaft of the on/off switch and potentiometer to pass through.**

    Be sure to wear safety glasses in case a piece of plastic flies off in the wrong direction!

LED1 from R1 to ground

R1 from +V to an open region

R5 from Pin 3 of IC1 to LED terminal block

C1 from Pin 2 of IC1 to ground

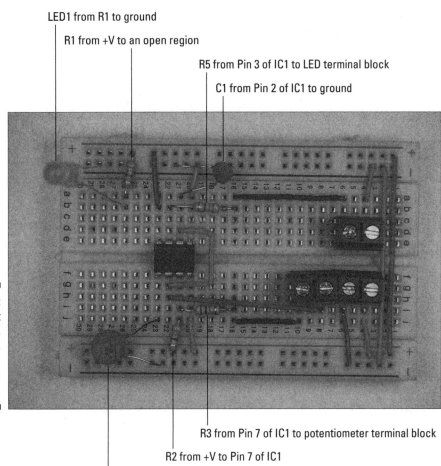

Figure 9-11:
Insert
resistors,
capacitors,
and an LED
on the
breadboard.

R3 from Pin 7 of IC1 to potentiometer terminal block

R2 from +V to Pin 7 of IC1

C2 from Pin 5 of IC1 to ground

14. **Attach the battery snap connector to the battery pack and place the breadboard in the pumpkin.**

    Because this pumpkin probably won't move around much, we just chose to lay the breadboard on the foam, as shown in Figure 9-15.

15. **Slip the on/off switch and the potentiometer through the holes you cut for them and use the nuts supplied with them to secure them, as shown in Figure 9-16.**

16. **Slip the LED socket through the hole you cut for it.**

    Figure 9-17 shows the LED socket installed in the pumpkin.

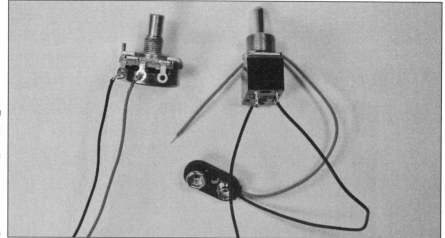

Figure 9-12:
Solder wires to the on/off switch and the potentiometer.

Figure 9-13:
An LED socket and a socket with LED inserted.

If the hole for the LED socket is small enough that the socket fits tightly, this "press fit" will hold it in place; if the socket is a little loose, use a little glue to secure it to the pumpkin. Read about what kind of glue to use in Chapter 3.

17. **Make sure that the on/off switch is in the off position and then attach the battery snap connector to a filled battery pack.**

18. **Place the battery pack in the pumpkin.**

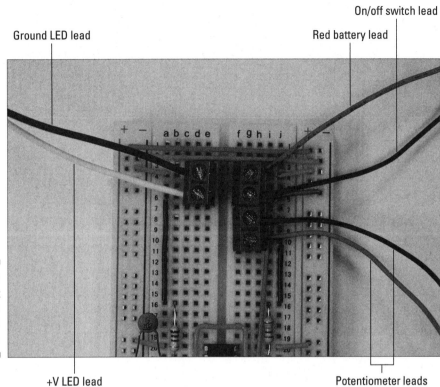

Ground LED lead

On/off switch lead

Red battery lead

Figure 9-14:
Connect
wires to the
terminal
blocks.

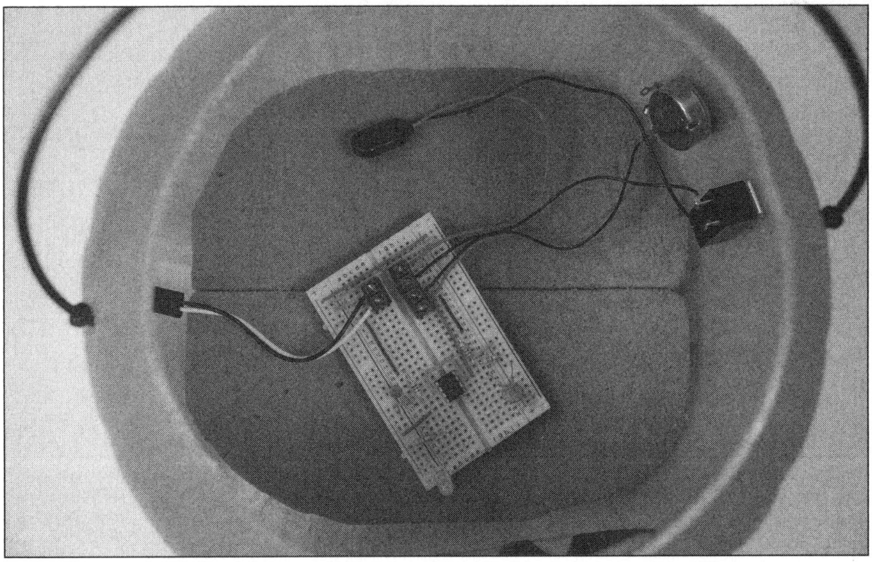

+V LED lead

Potentiometer leads

Figure 9-15:
The silent
pumpkin
with the
electronics
installed.

Figure 9-16:
Silent
pumpkin
with switch
and potenti-
ometer
installed.

Figure 9-17:
Silent
pumpkin
with LED
socket
installed.

Making a talking pumpkin

This is the mouthpiece of the pumpkin organization: the one that receives the IR beam and plays back whatever you record.

Follow these steps to build your talking pumpkin:

1. **Place the voice chip IC and six terminal blocks on the breadboard, as shown in Figure 9-18.**

   As you can see in this figure, you connect two wires to each terminal block. The wires from these six terminal blocks (TBs) go to the battery pack, on/off switch, record switch, IR detector, speaker, and microphone, respectively.

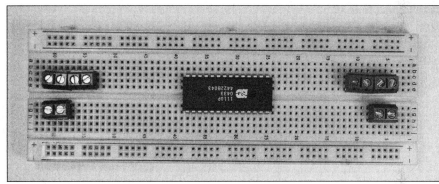

Figure 9-18:
Place the voice chip IC and six TBs on the breadboard.

2. **Insert wires to connect each component and terminal to the ground bus and insert a wire between the two ground buses to connect them, as shown in Figure 9-19.**

Figure 9-19:
Connect components to the ground bus.

In this figure, seven shorter wires connect components to ground bus (marked with a – on this breadboard); the long wire on the right connects the two ground buses together.

3. **Insert wires to connect each component and terminal to the +V bus and insert a wire between the two +V buses to connect them, as shown in Figure 9-20.**

   In this figure, five wires have been added: four shorter wires connect components to the +V bus (marked with a + on this breadboard); the long wire on the right connects the two +V buses.

Figure 9-20:
Connect components to the +V bus.

4. **Insert wires to connect the voice chip IC to the terminal blocks and to the open regions of the breadboard where discrete components will be inserted, as shown in Figure 9-21.**

5. **Insert discrete components on the breadboard, as shown in Figure 9-22.**

   The short lead of the LEDs and C2 go to the ground bus.

6. **Connect 6" wires (any color will work just fine) to the speaker and solder them, as shown in Figure 9-23.**

7. **Connect the black wire from the battery pack snap and another 6" black wire to the on/off switch and solder them together, as shown in Figure 9-23.**

8. **Connect a 6" red wire and a 6" black wire to the microphone pins, as indicated in Figure 9-24. Then solder them, as shown in upcoming Figure 9-25.**

9. **Connect a 6" red wire and two 6" black wires to the IR detector pins, as indicated in Figure 9-24; then solder them, as shown in upcoming Figure 9-25.**

   After the solder joints cool, wrap the microphone and IR detector solder joints with electrical tape to prevent them from coming into contact with each other.

10. **Connect two 10" wires (any color works) to the record switch and solder them, as shown in Figure 9-25.**

11. **Attach the wires from the battery pack snap connector, on/off switch, IR detector, microphone, and speaker to the terminal blocks, as shown in Figure 9-26.**

When attaching the wires to the terminal blocks, cut the wires to the length you need and strip the ends.

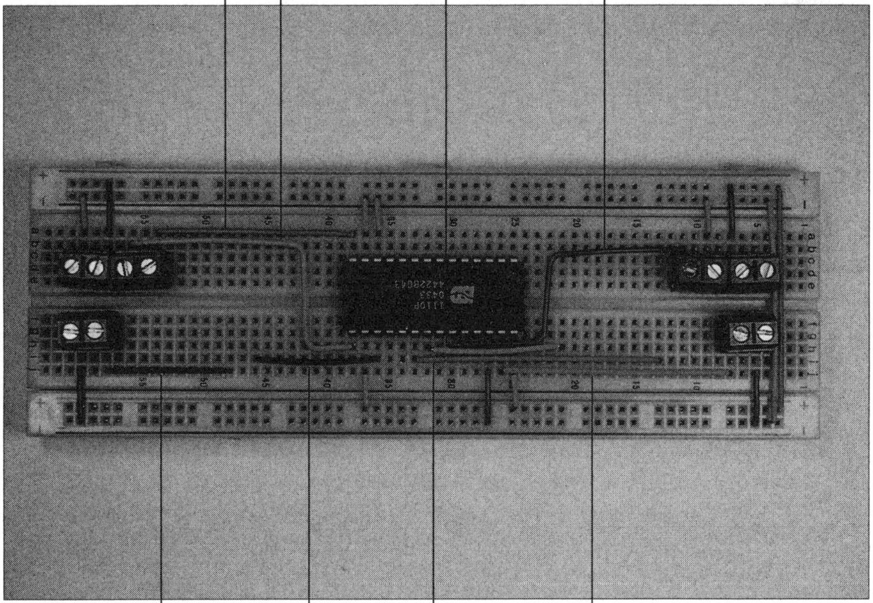

IC Pin 14 to terminal block for speaker

IC Pin 15 to terminal block for speaker

IC Pin 21 to open region of breadboard

IC Pin 23 to terminal block for IR detector

Figure 9-21: Connect the IC to the TBs and available locations for discrete components.

IC Pin 27 to terminal block for record switch

IC Pin 20 to open region of breadboard

IC Pin 17 to open region of breadboard

Open region of breadboard to terminal block for microphone

12. **Use your handy mini hacksaw or a utility knife to shape foam blocks to fit inside the bottom of the pumpkin.**

13. **On the side of the plastic pumpkin that will be out of sight, cut holes just large enough to allow the shaft of the on/off switch, microphone, and record switch to pass through.**

14. **On the side of the talking pumpkin that will face the silent pumpkin, cut a slot just large enough to allow the IR detector to fit through.**

Capacitor C1 and resistor R1 in series between wires to IC Pins 20 and 21

Resistor R4 between +V and an open region of the breadboard

LED2 between resistor R4 and ground

Resistor R2 between IC Pin 19 and ground

LED1 between IC Pin 14 and ground

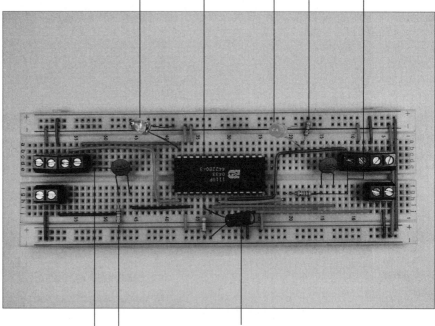

Figure 9-22:
Insert
discrete
components
on the
breadboard.

Capacitor C2 between IC Pin 19 and ground

Resistor R3 between capacitor C3 and +V

Capacitor C3 between microphone terminal block and IC Pin 17

15. **Place the breadboard and speaker on the foam. Then slip the on/off switch, microphone, and IR detector through the holes you cut, as shown in Figure 9-27.**

    Figure 9-28 shows the IR detector protruding from the side of the pumpkin.

16. **Tuck the wires off to the side; refer to Figure 9-27.**

    You can secure the on/off switch with the nut supplied and use either a press fit or a bit of glue to hold the microphone and IR detector in place.

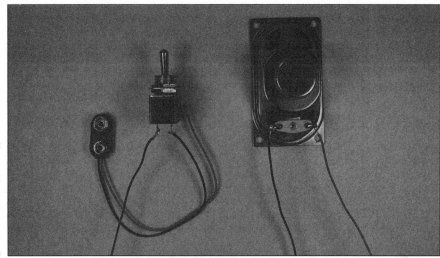

Figure 9-23:
Wires
soldered to
speaker and
on/off
switch.

17. **Slip the record switch through the hole you cut for it and secure it with the supplied nut; refer to Figure 9-27.**

    The direction of the threads on the body of the switch determines which way you insert it. The one we used had threads positioned such that the nut would be attached from inside the pumpkin, as shown from the outside in Figure 9-29.

18. **Insert the wires from the record switch to the remaining terminal block on the breadboard.**

19. **Make sure that the on/off switch is in the off position and attach the battery snap connector to a filled battery pack.**

20. **Place the battery pack in the pumpkin.**

    Your pumpkins are all ready to scare the pants off of your Halloween visitors. The completed talking pumpkin is shown in Figure 9-30.

Ground pin, use black wire

+V pin, use red wire

Figure 9-24:
Microphone
and IR
detector pin
identifi-
cation.

Vout pin, use black wire

Ground pin, use black wire

+V pin, use red wire

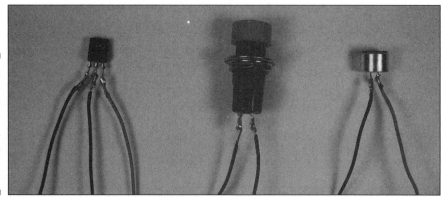

Figure 9-25:
Wires
soldered to
the
microphone,
IR detector,
and record
switch.

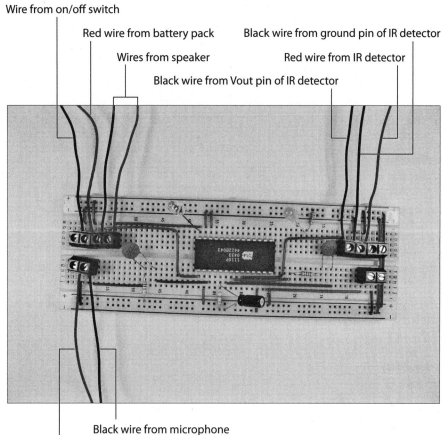

Wire from on/off switch

Red wire from battery pack

Black wire from ground pin of IR detector

Wires from speaker

Red wire from IR detector

Black wire from Vout pin of IR detector

Figure 9-26:
Attach
wires to
terminal
blocks.

Black wire from microphone

Red wire from microphone

Trying It Out

Those trick-or-treaters are probably lumbering up your street as we speak, so get your pumpkins set up and working to give them a proper scare.

Follow these steps to operate the scary pumpkins:

1. **Press the on/off switch of the talking pumpkin to turn it on.**

2. **Press and hold down the record switch and speak your greeting.**

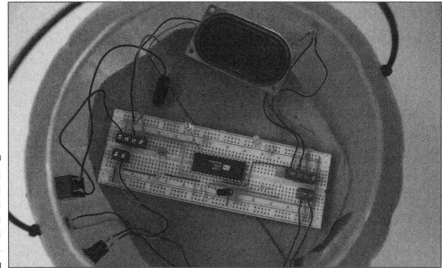

Figure 9-27:
Talking
pumpkin
with
electronics
in place.

Figure 9-28:
The IR
detector
mounted in
the side of
the
pumpkin.

3. Place the talking pumpkin on one side of your entryway (just inside, or safely out of the rain because these are not waterproof) and place the silent pumpkin on the other, with the IR LED facing the IR detector.

4. Flip the on/off switch of the silent pumpkin to On and adjust the potentiometer until the talking pumpkin starts talking.

5. Wait for your guests to walk between the pumpkins, and watch them scatter in fear (or laugh in delight) when the pumpkins sound off.

Figure 9-29:
Switches
and
microphone
mounted on
pumpkin.

Figure 9-30:
The talking
pumpkin, all
put
together.

Here the obvious things to check out if you're having a problem:

- ✔ All the batteries are fresh, are tight in the battery pack, and face the right direction.
- ✔ No bare wires near the microphone or IR detector touch each other.

If you need to troubleshoot your circuit beyond the obvious, here are some options:

- ✔ See whether the LEDs are inserted backward or are burned out.
- ✔ If the sound isn't loud enough, add an amplifier between IC1 and the speaker. For information about this procedure, see the application note on the manufacturer's Web site at `www.winbond-usa.com/products/isd_products/chipcorder/applicationbriefs/apbr06.pdf`.
- ✔ If nothing's shaking, make sure the IR detector wires are connected to the correct terminal block pins.
- ✔ Make sure that the IC isn't backward.

To test that the circuit in the silent pumpkin works, place a standard LED into the circuit instead of an IR LED. If you get a light, try a different IR LED or check the connection to the IR LED.

Taking It Further

Aren't talking pumpkins just so cool? (Sounds like a rap group, now that we think of it!) You can morph these guys into something else or expand their functionality in a few different ways:

- ✔ Obviously, you can change the containers for this light and sound event to whatever your heart desires. Plastic Santas, scarecrows, or firecrackers come to mind.
- ✔ You could have a one-two punch scenario for your pumpkins: One sound goes off when somebody steps into the beam, and another sound hits when somebody steps out of it. You need a receiver and transmitter in each pumpkin. Then connect the signal output of the IR detector to Pin 23 of the voice chip in a receiver circuit in one pumpkin, and Pin 24 of the voice chip in the receiver circuit of the other pumpkin.
- ✔ Try using a voice synthesizer chip. Instead of recording your own message, buy a voice synthesizer chip, such as the one we use in Chapter 7. This allows you to put together custom sound effects such as a rocket blasting off, a bit of Beethoven's Ninth Symphony, or a red alert alarm when somebody interrupts your IR beam.

Chapter 10

Dancing Dolphins

We all know that lighting effects make for a good time at movie premieres and parties. Building a design from lights and making the lights move in sequence can give you a great effect.

In this project, we show you how to create a series of dolphins dancing across the water (well, the plywood). You could just as easily create a series of spaceships in sequential stages of taking off, birds flying through the sky, or just about anything you can imagine.

Along the way, you can pick up some tips about timer chips, decade counters, and the artistic opportunity of stringing lots (and we do mean *lots*) of LEDs in sequence.

The Big Picture: Project Overview

When you complete this project, you'll have a wall display sporting five dolphins, outlined in LEDs, that light up one after the other, making them seem to dance across the wall. You can see the little guys jumping and jiving in the finished display shown in Figure 10-1.

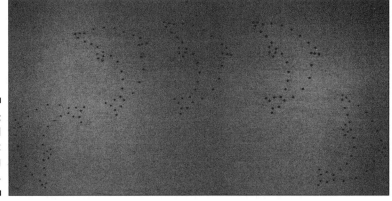

Figure 10-1:
The final
product:
dancing
dolphins.

Here's the big picture of the dancing dolphin project:

1. Put together an electronic circuit to control the timing of the light display.

2. Create a template for the dolphins and drill holes in a plywood sheet for five dolphin outlines by using the template.

3. Wire five arrays of LEDs and resistors onto the plywood sheet.

4. Mount the circuit on the plywood, connect the arrays of LEDs to the circuit, and add a plywood board to the back of the project.

5. Turn on the juice (that is, pop in the batteries).

   The circuit sends current to each group of LEDs for about two seconds, lighting up and then dimming the dolphins in sequence.

Scoping Out the Schematic

You have but one breadboard to put together for this project, but we make up for that by making you string 5 LED/resistor arrays, each containing 38 LEDs and 19 resistors.

Take a look at the schematic for the board, as shown in Figure 10-2.

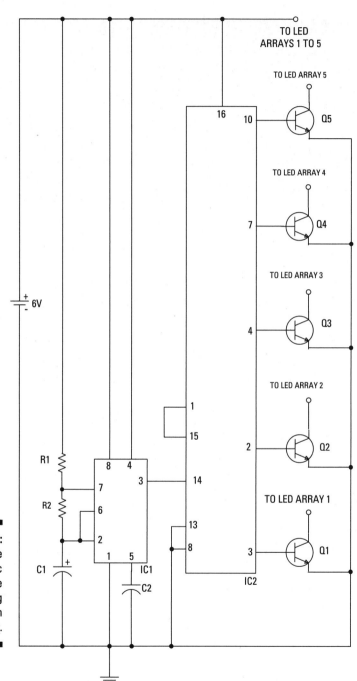

Figure 10-2:
The
schematic
of the
dancing
dolphin
circuit.

Getting in the swim: Exploring the dolphin circuit

To make your dolphin shapes light up in sequence, you need to make a circuit that uses a timer chip and a decade counter chip in combination with some resistors and a capacitor plus some transistors. Together, these control how often each of the five dolphins lights up and how long each stays lit.

A *decade counter* essentially takes a square wave and breaks it up into ten pulses. For those of you who took Latin, you'll recognize *decade* as related to the magic number ten. Read more about this counter in the following list.

Here's the overview of the schematic elements that you use to control your terpsichorean dolphins:

- ✔ **IC1** is a key component of this circuit; it's an LM555 timer chip that you use to generate a square wave at its output on Pin 3.

- ✔ **IC2** is the other key component of this circuit. This is a 4017 decade counter that takes a square wave and generates ten sequential pulse outputs. A 4017 decade counter does this by placing +V on one of its output pins at a time, one after the other. The 4017 decade counter switches to the next output pin at the start of each cycle of the square wave generated by the timer, as shown in Figure 10-3. This allows you to control the rate at which the 4017 decade counter switches +V to each output pin; this is done by controlling the frequency of the square wave generated by the LM555 timer chip. Because we didn't want ten dolphins, we connected the sixth output pin (Pin 1) to the reset pin (Pin 15). This applies +V to the reset pin after five dolphins dance across the wall and also resets the counter to the first output pin, skipping the last four output pins altogether.

- ✔ **R1, R2, and C1** are, respectively, two resistors and a capacitor that form the RC circuit that determines the frequency of the square wave generated by the LM555 timer chip.

- ✔ **Q1, Q2, Q3, Q4, and Q5** are 2N3053 transistors that turn on when the output pin of the 4017 decade counter they're connected to is switched to +V. You use these transistors to supply the necessary current — about 190 milliamps — to light the 38 LEDs in each group.

The 555 timer IC generates a square wave from its output. The frequency of the square wave that is generated is determined by how fast the capacitor fills and drains. You calculate how fast the capacitor fills to two-thirds of its capacity or drains to one-third of its capacity by using the *RC time constant equation*. You can read more about this equation in Chapter 9.

C2 is a capacitor that reduces the occurrence of noise on Pin 5 of the LM555, which could cause false triggering of the IC. This noise can occur if Pin 5 is left unconnected (also called *floating*).

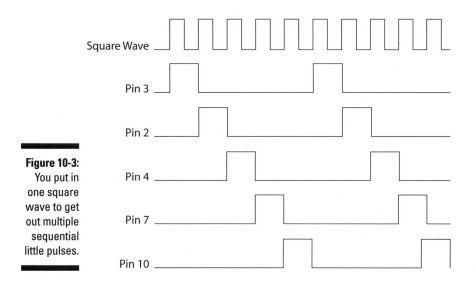

Figure 10-3:
You put in
one square
wave to get
out multiple
sequential
little pulses.

Setting up the light show

The circuit won't mean a thing if you don't set up the lights for it to control. That's where the elements of the LED/resistor arrays come in. An array, in this case, equals the lights that define one whole dolphin.

These five arrays each include

> **38 LEDs,** which light up when a river of current runs through them

> **19 resistors,** which are resistors that limit the current running through the LEDs in series with each resistor to approximately 10 milliamps

Take a look at the schematic for these in Figure 10-4. Notice that we have not assigned a number to each LED and resistor. That is because we have 190 LEDs and 95 resistors among the 5 dolphins — and we don't have time.

These LEDs and resistors are wired together such that two LEDs and one resistor are in series. When +V is applied, current runs through each component sequentially. Each group of two LEDs and one resistor are connected parallel with the other groups of LEDs and resistor. Because 19 LED/resistor groups are in each array, the total current running through an array is approximately 190 milliamps.

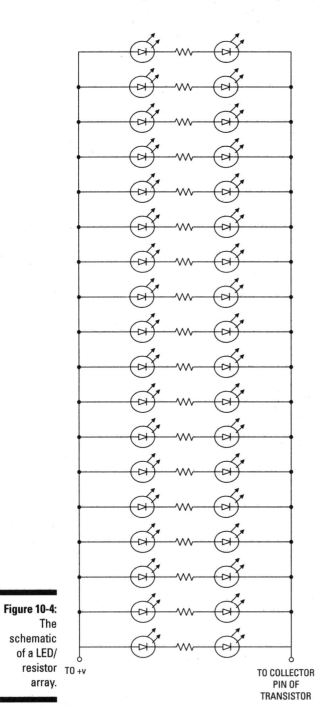

Figure 10-4:
The
schematic
of a LED/
resistor
array.

TO +v

TO COLLECTOR
PIN OF
TRANSISTOR

Building Alert: Construction Issues

The back of the dolphins — where all the resistors and LED leads are soldered together — gets kind of hectic. Make sure that if a lead gets bent in the course of putting it on the plywood that it doesn't short. Rather than using electrical tape, we used liquid electrical tape to coat the exposed leads.

For a mounting surface, we used ¼" sheet plywood as the best bet. We tried using PVC (a kind of plastic) sheets, but they weren't rigid enough to support the display.

We won't kid you: Stringing the LED and resistor arrays takes some time and patience. There is no real shortcut to offer you here! We recommend that you solder the resistors to the LED leads first and then attach and solder either the red or black wires. If you try to attach all the wires before you solder, some will get in the way of your soldering iron.

You can use any artwork you like to create the images of your choice. We found a simple clip art drawing of a dolphin, printed it, copied and enlarged it, and used that as our stencil. We tipped the stencil at different angles across the plywood, creating a feeling of movement. However, you can use any graphic for your template as long as you avoid using anything so complex that people can't make it out by looking at a simple outline of lights.

Perusing the Parts List

It's time to go shopping for those electronic parts you use to build the circuits and assemble all those LEDs into dancing dolphins. This is where we give you the parts lists for the circuit and the LED arrays.

A circuit with a porpoise

The circuit that controls the timing of your light show involves the following parts, several of which are shown in Figure 10-5:

- ✔ **47 kohm resistor (R1)**
- ✔ **470 kohm resistor (R2)**
- ✔ **LM555 timer (IC1)**
- ✔ **4017 decade counter (IC2)**
- ✔ **1 microfarad electrolytic capacitor (C1)**

- ✓ 0.1 microfarad ceramic capacitor (C2)
- ✓ Five 2N3053 transistors (Q1, Q2, Q3, Q4, Q5)
- ✓ Breadboard (830 contacts)
- ✓ Six 2-pin terminal blocks
- ✓ An assortment of different lengths of prestripped short 22 AWG wire

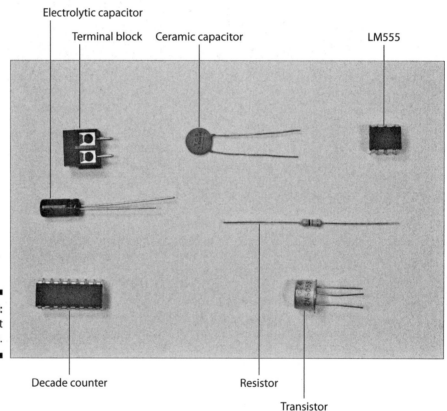

Electrolytic capacitor

Terminal block   Ceramic capacitor

LM555

Figure 10-5:
Key circuit
components.

Decade counter

Resistor

Transistor

Making your dolphins boogie

Here's your shopping list for building your dancing dolphin light display. This part of the project involves the following parts, several of which are shown in Figure 10-6:

- ✓ 190 orange T-1 ¾ LEDs
- ✓ 95 220 ohm resistors
- ✓ Six ⅝" 6-32 flathead screws

✔ Six 2" standoffs with 6-32 threads

✔ Several feet of black 20 AWG wire

✔ Several feet of red 20 AWG wire

✔ A four pack of AA batteries with an on/off switch

✔ An assortment of different lengths of prestripped short 22 AWG wire

✔ Seven wire clips

✔ Two sheets of ¼" plywood measuring 2' high x 4' wide

✔ Liquid electrical tape

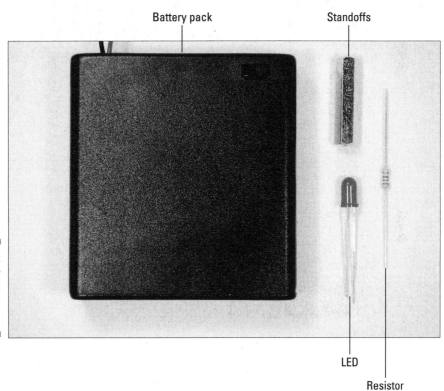

Figure 10-6:
Key
components
of the LED
arrays.

Taking Things Step by Step

To make a dolphin dance a two-step, you have to do a few things. First, you have to build the circuit that makes it all run. Then you have to assemble the lights that outline the dolphins on a surface, such as plywood. That's what we cover in this section.

Making the circuit

Time to go one-on-one with your breadboard. Here are the steps involved:

1. **Place the LM555 IC, the LM4017 IC, and six terminal blocks on the breadboard, as shown in Figure 10-7.**

   The six terminal blocks in this figure will be used to connect two wires each to various components in the circuit. The wires from one terminal block go to the battery pack, and the wires from each of the other terminal blocks go to one of the LED/resistor arrays.

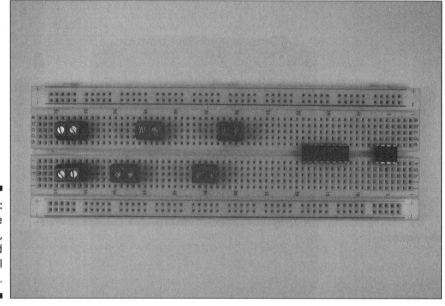

Figure 10-7:
Place the
LM555 IC,
4017 IC, and
six terminal
blocks.

2. **Insert discrete components on the breadboard, as shown in Figure 10-8.**

C2 from Pin 5 of IC1 to ground

R1 from an open region to +V

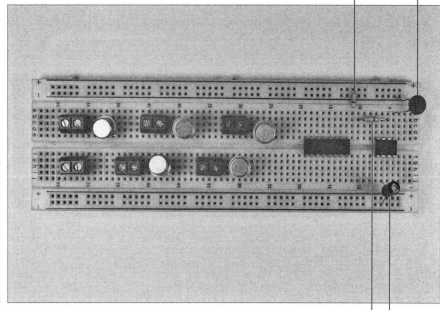

Figure 10-8:
Insert
resistors,
capacitors,
and
transistors
on the
breadboard.

R2 from Pin 6 of IC1 to R1

C1 from Pin 2 of IC1 to ground

Figure 10-9 identifies the pins of the 2N3053 transistors.

You insert each transistor pin into a separate breadboard row, placing the collector pin nearest to the terminal block, and then insert C1 with the shorter of the two pins of C1 in the ground bus and the longer pin in the same breadboard row as Pin 5 of IC1.

Base

Emitter   Collector

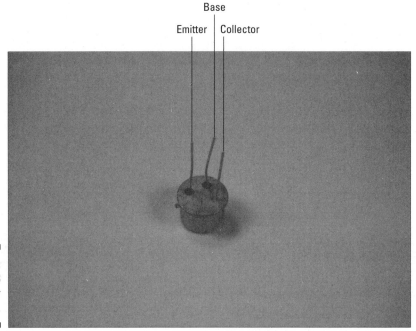

Figure 10-9:
The 2N3053
transistor
pinout.

3. **Insert wires to connect the ICs, the battery pack terminal block, and the emitter pin of the transistors to the ground bus. Then insert a wire between the two ground buses to connect them, as shown in Figure 10-10.**

Figure 10-10:
Nine shorter wires connect components to ground bus; the long wire on the left connects the two ground buses.

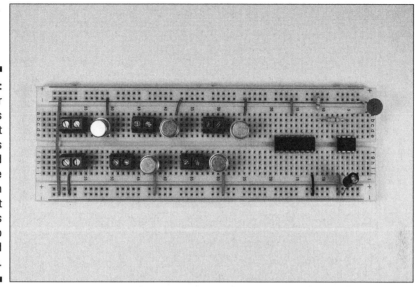

4. **Insert wires to connect the ICs and the terminal blocks to the +V bus, as shown in Figure 10-11.**

Pin 8 of IC1 to +V

Terminal blocks to +V   Pin 16 of IC2 to +V

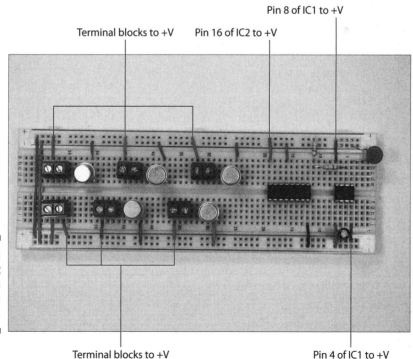

Figure 10-11:
Connect components to the +V bus.

Terminal blocks to +V   Pin 4 of IC1 to +V

5. **Insert wires to connect the collector pin of the transistors to the terminal blocks, as shown in Figure 10-12.**

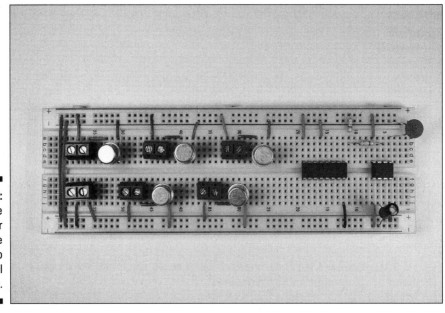

Figure 10-12:
Connect the collector pin of the transistor to terminal blocks.

6. **Insert wires to connect the ICs, terminal blocks, and discrete components, as shown in Figure 10-13.**

Pin 2 of IC1 to Pin 6 of IC1

Pin 7 of IC1 to R1

Pin 10 of IC2 to base pin of Q5

Pin 1 of IC2 to Pin 15 of IC2

Pin 4 of IC2 to base pin of Q3

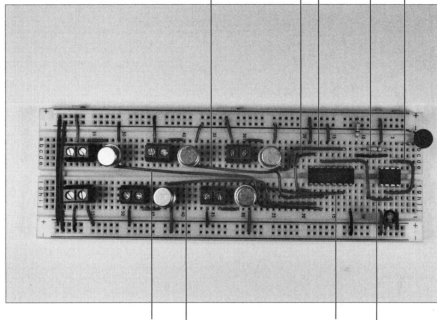

Figure 10-13:
Hook up the
IC, terminal
blocks, and
discrete
components.

Pin 3 of IC2 to base pin of Q1

Pin 2 of IC2 to base pin of Q2

Pin 7 of IC2 to base pin of Q4

Pin 3 of IC1 to Pin 14 of IC2

Making dolphins

All the brains of the circuit assembled in the previous section are there to make the dolphin light display work.

Follow these steps to create your dancing dolphin display:

1. **Make five dolphin (or another figure) stencils, each about 11" high.**

   You can create a template by printing a piece of clip art or other simple drawing and enlarging it with a copier.

2. **Decide where you want to place the dolphins on the plywood and use double-sided tape to affix the templates.**

3. **Use a marking pen to mark where you want to place the LEDs on the plywood to show the outline of the dolphins.**

   We used 38 LEDs per dolphin. We spaced the marks about 1½" apart in the parts of the dolphin where there was little change in shape. Where the dolphin's shape was a bit more complex (for example, the nose and tail), we spaced them more closely.

4. **Drill test holes in a piece of scrap wood to determine the size of drill bit that you should use to give a press fit for the LEDs.**

   We used a ¹³⁄₆₄" drill bit.

5. **Drill holes for the LEDs at the locations that you marked in Step 3.**

   The plywood after drilling is shown in Figures 10-14 and 10-15.

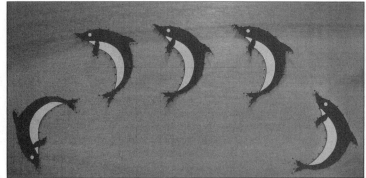

Figure 10-14:
The plywood after drilling holes.

Figure 10-15:
A closer look at the dolphin templates with all holes drilled.

6. **Pick a dolphin and start inserting LEDs in the drilled holes.**

   We suggest starting at either end of the plywood sheet so the LED leads of a finished dolphin aren't in your way while you work.

7. **Attach resistors between every other LED, as shown in Figure 10-16.**

   Attach the resistors to the short lead on the first LED of each pair and to the long lead on the second LED of each pair. At this point, leave the long lead on the first LED and the short lead on second LED alone.

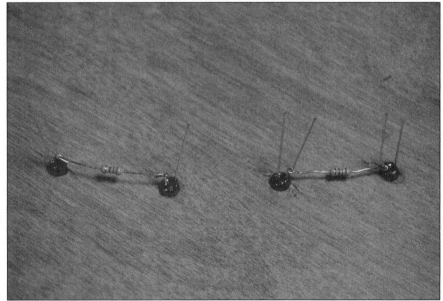

Figure 10-16:
Attach
resistors
to LEDs.

8. **Solder the resistors to the leads and clip the leads just above the solder joint.**

   Clip only the leads to which you have soldered resistors. Figure 10-17 shows how the dolphin board should look at this point.

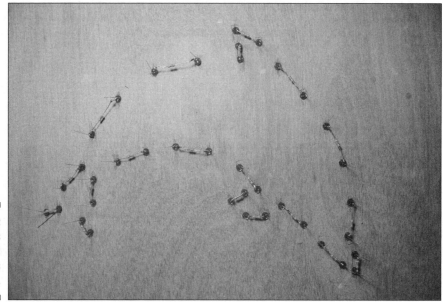

Be sure to heed all the safety precautions about soldering that we provide in Chapter 2. For example, don't leave your soldering iron on if you have to step away. And just in case a bit of solder has an air pocket that could cause it to pop, wear your safety glasses whenever you solder.

9. **Connect the short leads on every other LED to short lengths of 20 gauge black wire, as shown in Figures 10-18 and 10-19.**

   Figure 10-18 shows a gap without a black wire across the tail of the dolphin. One of the LEDs at the tail with a short lead has only one black wire attached, and the other has a 2' 20 gauge black wire attached; you connect this wire to a terminal block on the breadboard in Step 20.

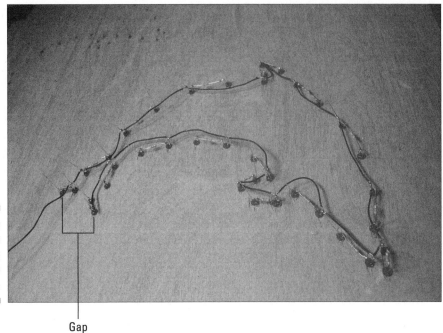

Figure 10-18:
Black wires
connecting
short leads
of LEDs.

Gap

10. **Solder the black wires to the leads, as shown in Figure 10-19.**

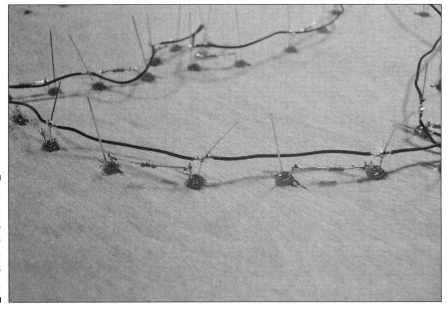

Figure 10-19:
A close-up
of black
wires
connecting
short leads
of LEDs.

11. **Connect the long leads on every other LED to short lengths of 20 gauge red wire, as shown in Figures 10-20 and 10-21.**

    Figure 10-20 shows a gap where no red wire is strung across the tail of the dolphin. One of the LEDs at the tail with a long lead has only one red wire attached; the other has a 2' 20 gauge red wire attached, which you connect to a terminal block on the breadboard in upcoming Step 20.

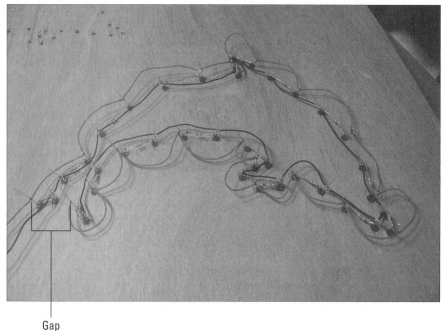

Figure 10-20: Red wires connect long leads of LEDs.

Gap

12. **Solder the red wires to the leads, as shown in Figure 10-21.**

13. **Clip the LED leads just above the solder joint.**

14. **Make sure that the LED leads and solder joints don't touch each other and then coat them with liquid electrical tape to help prevent any shorts if wires are bent or pushed together.**

15. **Repeat Steps 6–11 for each dolphin until you install and wire the LEDs for all five.**

16. **Chose a location on the plywood sheet to place the battery pack so that you can reach the on/off switch.**

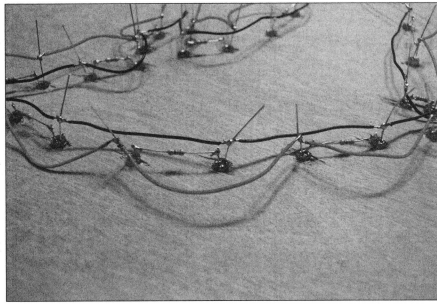

Figure 10-21:
A close-up
of red wires
connecting
long leads
of LEDs.

17. Attach Velcro to the battery pack and the plywood so that you can attach the battery pack to the plywood, as shown in Figure 10-22.

Figure 10-22:
The battery
pack in
place.

18. **Chose a location on the plywood sheet where you will place the breadboard.**

19. **Attach Velcro to the breadboard and the plywood and then attach the breadboard to the plywood, as shown in Figure 10-23.**

20. **Insert the wires from the battery pack and the dolphins to the terminal blocks on the breadboard, as shown in Figure 10-23.**

If the wire on the battery pack isn't long enough to reach the breadboard, splice and solder longer 20 gauge red and black wires. Protect the splices with electrical tape or heat shrink tubing.

Red wire from dolphin 1

Black wire from dolphin 1

Red wire from dolphin 3

Black wire from dolphin 3

Red wire from dolphin 5

Black wire from dolphin 5

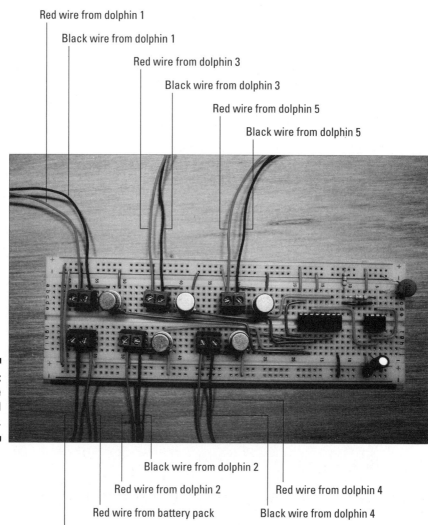

Figure 10-23:
The breadboard in place.

Black wire from dolphin 2

Red wire from dolphin 2

Red wire from battery pack

Red wire from dolphin 4

Black wire from dolphin 4

Black wire from battery pack

21. **Secure the wires with wire clips.**

22. **Add the protective backing by performing the following steps:**

   a. *Drill holes for 6-32 screws in six locations on each plywood sheet to attach the standoffs between the two plywood sheets.*

   b. *Secure the six standoffs to the plywood sheet on which you've placed the LEDs, using 6-32 screws.*

   c. *To finish off the project, secure the second plywood sheet to the six standoffs using 6-32 screws, as shown Figure 10-24.*

Figure 10-24:
The finished
dancing
dolphin
product.

Trying It Out

We're sure that after spending hours stringing LEDs, you're eager to see the fruits of your labors. (We were!) Whether you followed our lead and created dolphins or got creative with some other shape, it's time to turn on your lighting display.

You can either lean the display against a wall or hang it as you would a large picture; then enjoy the show at your next dance party.

There aren't many steps to getting this project going:

1. **Pop the batteries into the battery pack.**
2. **Flip the switch to On.**

That's it! Your dolphins will begin dancing across the wall in timed sequence.

Here are the obvious things to check out if you're having a problem:

- ✔ All the batteries are fresh and tight in the battery pack and also face the right direction.
- ✔ If one dolphin in the group doesn't function, check its wiring.
- ✔ If one or two LEDs aren't working, replace them.
- ✔ If two LEDs in series with each other aren't functioning, you might have reversed the long and short leads of the LEDs. If so, it's easiest to just replace that pair.

Taking It Further

Aren't dancing dolphins just so cool? You can morph these guys into something else or expand their functionality in a few different ways:

- ✔ **Change the stencils to create whatever your heart desires.**

    Santa and his reindeer, or swans, or leaping lizards come to mind.
- ✔ **You can create larger figures or make up to ten figures.**

    However, we wouldn't suggest using more LEDs than we used here to outline them because that would drain the batteries too quickly or possibly cause the 2N3053 transistors to overheat.
- ✔ **Get your dolphins chattering in sequence.**

    You could add sound by using a sound chip, as we show you in Chapter 14.
- ✔ **Use a SpeakJet sound synthesizer chip (like we show you in Chapter 7) to have each dolphin make a unique sound, giving each a distinctive personality.**

    Tie an output pin of the decade counter to an event pin on the SpeakJet sound synthesizer chip in parallel with the transistor for the LEDs. Program the SpeakJet to trigger an event when it goes from low to high. When a dolphin lights up, it triggers the sound for that pin.

Chapter 11

Controlling a Go-Kart, Infrared Style

In This Chapter

▶ Looking over the go-kart schematic

▶ Laying down the parts list

▶ Breadboarding the transmitter and receiver

▶ Building the go-kart body

▶ Making the go-kart go!

Remote-controlled vehicles are very cool things. You can cause them to zip around your living room, tease your cat, and race with your friends. Building your own go-kart from scratch and using the power of infrared to control its every move are what this project is all about. How far you go in making a cool-looking car is up to you; here we've designed what we call an *infrared* go-kart — sort of like a VW Bug but just a little bit smaller.

In this chapter, you get to explore creating a go-kart that can go forward and backward and even turn left and right on a dime. Along the way, you pick up all kinds of tips about infrared transmitters and receivers as well as controlling motors.

The Big Picture: Project Overview

After you complete this project, you'll have a little three-wheeled go-kart that you can control with an infrared transmitter, as shown in Figure 11-1. Here's the rundown on the key features of this vehicle:

✔ **We use three wheels.** Why three wheels? No kart you ever saw had three wheels, right? Well, the world of electronic projects isn't Detroit, so we use three wheels because it's simpler and it works. As soon as you use four wheels, you have to add a suspension system to ensure that all the wheels stay in contact with the ground, especially when that ground gets uneven. For a simple electronics project, where the focus is on the electronics and not the mechanics, you don't want to deal with a suspension system.

✔ **The infrared remote control works along the same lines as your television remote control.** You supply an electric current that causes the LED to generate infrared light. The IC in the transmitter modulates the electric current running through the LED according to which button you push on the transmitter: on/off, motor right (we'll call this *motor R*), or motor left *(motor L)*. If you aim the transmitter at the infrared detector on the kart, the detector turns the infrared light back into electric current. That current is then interpreted by the receiver circuit on the kart to turn the motors on or off, reverse motor R, or reverse motor L (based on which button you push).

Figure 11-1:
The final product: a three-wheeled, infrared-controlled go-kart.

So what, exactly, will you be doing in this project? The project involves

1. Putting together the electronic circuit for the IR transmitter and fitting it into a plastic box with buttons that you use to control the movement of the go-kart

2. Putting together the electronic circuit for the IR receiver/motor driver

3. Building the base for the go-kart itself and attaching the various bits to it

Scoping Out the Schematic

There are actually two breadboards to assemble in this project: a transmitter circuit and a receiver circuit. Take a look at the schematics for these two in the following sections, along with helpful tips for reading them.

Transmitting at the speed of light

The transmitter is what you use to operate the go-kart. The transmitter circuit is shown in Figure 11-2.

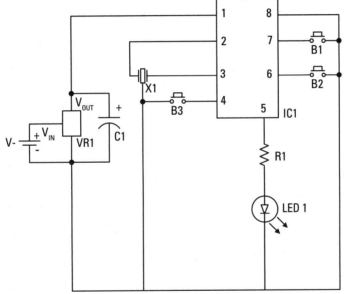

Figure 11-2:
The schematic of the transmitter circuit.

Here's the gist of what's going on:

- ✔ **VR1** is a voltage regulator that takes the 6 volts supplied by the battery pack and converts it to 5 volts, the maximum voltage specified for the decoder (IC1).

- ✔ The capacitor **(C1)** that's placed between the output pin and the ground pin of the voltage regulator prevents any *oscillation* (wiggling around) in the output voltage of the regulator.

- ✔ **IC1** is an encoder whose job it is to send out a signal that tells a decoder on the receiver what to do. Pins 4, 6, and 7 are inputs to the encoder (inputs 3, 2, and 1, respectively). When a normally open (NO) pushbutton switch **(B1, B2, B3)** tied to one of these pins is closed, the encoder modulates the 38 kHz carrier wave that tells the decoder IC in the receiver exactly which button has been pressed. This signal goes out through Pin 5 and then through a 150 ohm resistor **(R1)**. The resistor limits the current to about 22 milliamps; that's so you don't burn out the LED. The signal then goes through the IR LED, which generates an infrared signal.

- ✔ **X1** is a 4 MHz ceramic resonator. This works with components within the encoder to generate timing signals that help generate the 38 kHz carrier wave and runs an internal clock used to generate a signal that identifies which pushbutton switch has been closed.

Receiving what the transmitter sends

Just like your TV receives the signal from a remote control telling it to flip over to MTV, something has to receive the transmitter signal to make the kart go. The receiver circuit is shown in Figure 11-3.

Note what's going on in this schematic:

- ✔ The **IR detector** contains a photodiode; when the infrared signal reaches the photodiode, it produces an electrical signal. This electrical signal goes into Pin 4 of the decoder **(IC1)**.

- ✔ IC1 then decodes the electrical signal sent by the transmitter. Pressing and releasing a pushbutton on the transmitter causes the decoder to switch around the state of the corresponding output; Pin 7 is output 1, Pin 6 is output 2, and Pin 5 is output 3. So, if the output pin is high, (5 volts), it will be changed to ground; if the output pin is low (ground), it will be changed to high (5 volts). The output stays at that voltage until another signal appears to change it.

- ✔ The resonator **(X1)** drives an oscillator within the decoder to generate internal clock signals used to decode the signal sent out by the encoder.

- ✔ Like with the transmitter circuit, a voltage regulator **(VR1)** limits the supply voltage to the ICs to 5 volts.

    S1 is the power switch for the receiver.

- ✔ You place a 10 microfarad capacitor **(C2, C4, C6)** and 0.1 microfarad capacitor **(C1, C3, C5)** between the +V and ground buses at the +V input of each of the ICs. These are used to filter electrical noise from the DC motors; that noise can prevent the ICs from operating correctly because they won't consistently have the correct supply voltage. If you don't add these capacitors to the circuit, the motors could occasionally stop running or stop responding when you press the transmitter button.

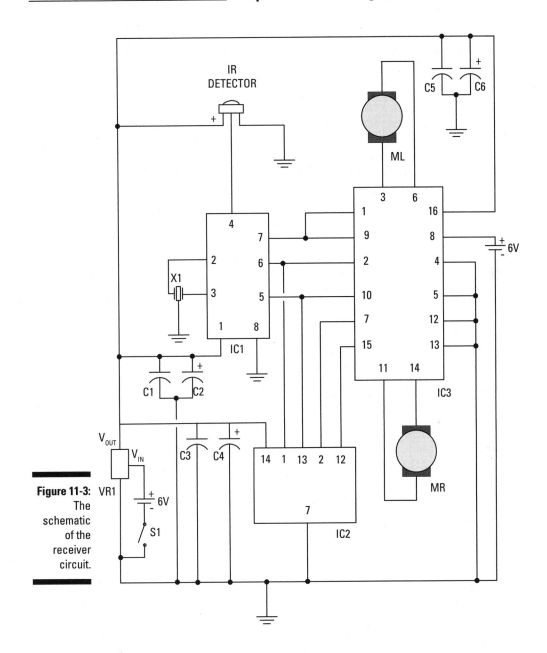

Figure 11-3: The schematic of the receiver circuit.

Controlling motor behavior

To control the movement of the kart, you need to be able to control the direction of both DC motors. Here's what will happen, depending on what the motors are doing:

✔ **Both motors are rotating forward:** The kart will move forward.

✔ **Both motors are rotating backward:** The kart will move backward.

✔ **If motor L is moving backward and motor R is moving forward:** The kart will make a left turn.

✔ **If motor L is moving forward and motor R is moving backward:** The kart will make a right turn.

In order to change the direction of the motors, you need to change the direction of the current flowing through the motor windings. A circuit called an *H-bridge* flips the direction of the current through the motors. In the receiver schematic (refer to Figure 11-3), IC3 contains two H-bridges. (You can't see the H-bridges in the schematic, but trust us: They're in there. You'll see where they go when you get to the steps for assembling the go-kart.)

To control the motors, the H-bridge needs the following inputs, as shown in Figure 11-3:

✔ **Pin 16:** +V to power the IC.

✔ **Pin 8:** A separate +V to power the DC motors.

This is why you need to use a second battery pack; it helps isolate the ICs from electrical noise generated by the DC motors that can cause the circuit to stop functioning intermittently. Read about this in the upcoming parts list section for this project.

✔ **Pins 1 and 9:** The signal, supplied from Pin 7 of the decoder (IC1), goes to Pins 1 and 9 of the H-bridge in IC3. If Pin 7 (IC1) is at +V, both DC motors turn on. If Pin 7 of IC1 is at ground, both DC motors turn off.

✔ **Pins 2 and 7:** Pins 2 and 7 of the H-bridge determine in what direction motor L rotates. To rotate motor L, the H-bridge requires +V at Pin 2 and ground at Pin 7 in order to go in one direction, or the opposite to go in the other direction. In order to establish the +V and ground connections for motor L, you connect Pin 6 of the decoder to Pin 2 of the H-bridge and also connect Pin 6 to IC3. IC2 inverts the signal so that +V becomes ground or ground becomes +V. You then connect the inverted signal to Pin 7 of the H-bridge; this gives you both the +V and ground you need to control the direction of motor L.

✔ **Pins 10 and 15:** These control the direction of motor R in the same way that Pins 2 and 7 control motor L. To rotate motor R, the H-bridge requires +V at Pin 10 and ground at Pin 15, or the opposite to go the other way. Connect Pin 5 of IC1 to Pin 10 of the H-bridge and also connect Pin 5 of IC1 to IC2. IC2 then inverts the signal so that +V becomes ground or ground becomes +V. That signal now goes to Pin 15 of the H-bridge.

Building Alert: Construction Issues

We thought and thought about what to use to place over the kart's works to give our kart a top. In a stroke of genius (well, we couldn't think of anything else, to be honest, and we were grocery shopping at the time), we used a plastic food storage container to create a bubble-like dome. It's cheap, easy to work with, and clear so you can see what makes the kart go, which is kind of cool. Make sure you get a container made of flexible plastic. Flexible plastic is easier to use because you have to make two cuts in this container: an opening in the back that lets the IR signal from your transmitter reach the IR detector in the back of the kart and an opening in the side so you can reach the power switch.

You can build the base of the kart out of ¼" thick plywood or ¼" rigid expanded PVC (plastic). The PVC provides a better finished look than the plywood, but either will work. You should be able to locate ¼" thick plywood at any lumber or homebuilding store.

A couple of robot supply houses, such as Budget Robotics and Solarbotics (here the material is called Sintra), sell small sheets of rigid expanded PVC. A Google search for **"rigid expanded PVC"** will also turn up plastic supply companies that sell larger sheets. Find more at these Web sites: www.budget robotics.com and www.solarbotics.com.

If you don't feel ambitious, you can leave off the bubble top and let the go-kart be a convertible model. If you plan to reuse the components after trying out the kart, that might be the way to go. However, if you plan to keep the kart intact, we suggest that you use a top to keep out dirt and avoid the possibility of wires or components being knocked off (in case your kart gets into a traffic accident in your living room).

The motor lugs used in this project are made of thin metal and will break off if you put too much stress on them. By using stranded wire, rather than solid wire, you can minimize the stress on the lugs.

Perusing the Parts List

We broke the parts list for this project into two very logical sections: one for the transmitter, and one for the receiver and the kart body on which the receiver is mounted.

Go-kart transmitter parts list

The transmitter (see its schematic in Figure 11-2) includes the following parts (several of which are shown in Figure 11-4):

- **150 ohm resistor (R1)**
- **Infrared encoder (IC1)**

  We use a Reynolds Electronics Tiny-IR Encoder IC because it requires less circuitry than similar ICs. Optional ICs are listed in the upcoming section, "Taking It Further."

- **5 volt regulator (VR1) (LP2950 5 volt regulator or similar)**
- **4.7 microfarad electrolytic capacitor (C1)**
- **IR LED, TSAL7200 (LED1)**

  Note that various other IR LEDs will work; www.rentron.com has a useful listing of similar IR LEDs on its site.

- **3-pin, 4 MHz ceramic resonator (X1)**

  2-pin resonators look an awful lot like 3-pin resonators, so be careful when you're ordering from one of those catalog: Read the small print so that you get the right one for this project.

- **Breadboard**
- **LED panel mount socket (size T-1 ¾)**
- **Four pack of AA batteries with snap connector**
- **Five 2-pin terminal blocks**
- **Enclosure (RadioShack part #270-1806 or similar)**
- **Velcro**
- **An assortment of different lengths of prestripped short 22 AWG wire**

  You can cut and strip the wire yourself, but for short lengths, we find it much easier to use the prestripped wire. If you have any kind of a life at all, spending time endlessly stripping wires just isn't worth your while!

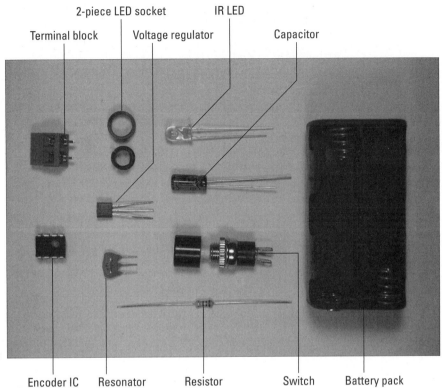

2-piece LED socket   IR LED

Terminal block   Voltage regulator   Capacitor

Figure 11-4:
Key
transmitter
components.

Encoder IC   Resonator   Resistor   Switch   Battery pack

Go-kart receiver/chassis parts list

The receiver and go-kart chassis (see the schematic in Figure 11-3) use the following parts (several of which are shown in Figures 11-5 and 11-6):

- ✔ **IR detector PNA4602M**

  Various other IR detectors will work. If you want to try another, you can find a listing of IR detectors at www.rentron.com.

- ✔ **Decoder (IC1) (Reynolds Tiny-IR Decoder)**
- ✔ **5 volt regulator (VR1) (LM7805 5 volt regulator or similar)**
- ✔ **Inverter (IC2) (SN74F04 or similar)**
- ✔ **H-bridge (IC3) (L293D or similar)**
- ✔ **3 pin, 4 MHz ceramic resonator (X1)**
- ✔ **Three 10 microfarad electrolytic capacitors (C1, C3, C5)**

- ✔ **Three 0.1 microfarad ceramic capacitors (C2, C4, C6)**
- ✔ **Four 2 pin terminal blocks**
- ✔ **Two 4 packs of AA batteries**
- ✔ **Two DC gear motors (MR, ML) (GM2 with 2⅝" wheels or equivalent)**

   We use the GM2 model gear motors because the suppliers (Hobby Engineering, www.hobbyengineering.com; and Solarbotics) carry wheels just made to fit them, and they cost a bit less than *servomotors,* a kind of motor often used in robots because it offers more control of quick changes in motion.

- ✔ **2" swiveling castor**
- ✔ **SPST (single-pole, single-throw) switch (S1)**
- ✔ **Two wire clips**

   Use RadioShack part #278-1668 or something similar; essentially, you can use anything that will secure the wires without damaging them.

- ✔ **¼" thick expanded PVC, 9" x 6½".**

   You could also use ¼" thick plywood.

- ✔ **6" x 9" plastic food container**
- ✔ **Velcro**

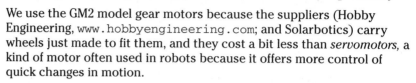

Figure 11-5: Key receiver components.

Voltage regulator 0.1 microfarad capacitor Terminal block H-bridge IC Inverter IC

IR detector   Resonator   10 microfarad capacitor   Decoder IC

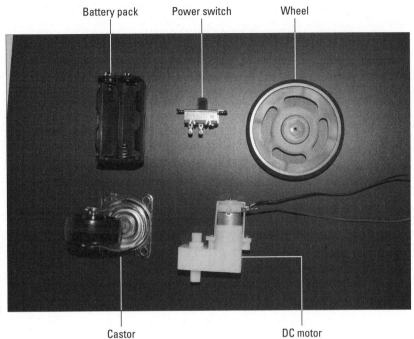

Battery pack   Power switch   Wheel

Figure 11-6:
Key go-kart
chassis
components.

Castor   DC motor

Taking Things Step by Step

Creating your go-kart involves building the transmitter, building the receiver, and then assembling the go-kart body with its motors and three wheels. The receiver circuit gets mounted on the body as well, and then you top the whole thing off with the plastic bubble cover.

Making the transmitter

The transmitter — the thing that you hold in your hand like a TV remote control — is what you aim at the go-kart to control it.

Follow these steps to build the transmitter circuit:

1. **Place the encoder IC, resonator, voltage regulator, resistor, and capacitor on the breadboard, as shown in Figure 11-7.**

   Make sure you put the negative lead of the capacitor in the same breadboard row as the center pin of the voltage regulator and the positive lead in the same breadboard row as the +5 volt output pin of the regulator.

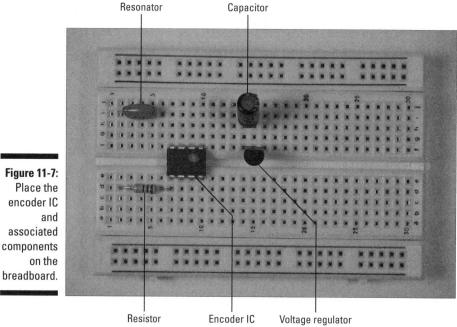

Resonator   Capacitor

Figure 11-7:
Place the encoder IC and associated components on the breadboard.

Resistor   Encoder IC   Voltage regulator

Figure 11-8 shows the uses of each pin of the voltage regulator, capacitor, and LED.

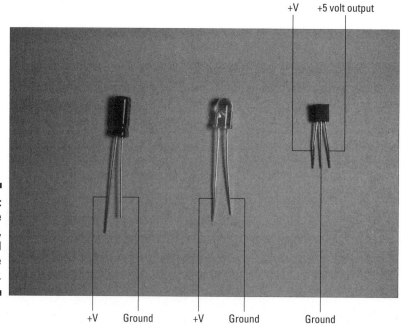

+V   +5 volt output

Figure 11-8:
Pinout of the capacitor, LED, and voltage regulator.

+V   Ground   +V   Ground   Ground

2. **Place five terminal blocks on the breadboard, as shown in Figure 11-9.**

   The five terminal bocks shown in this figure will be used to connect two wires each to various components in the circuit. The wires go to the battery pack, IR LED, on/off switch, motor R switch, and motor L switch, respectively.

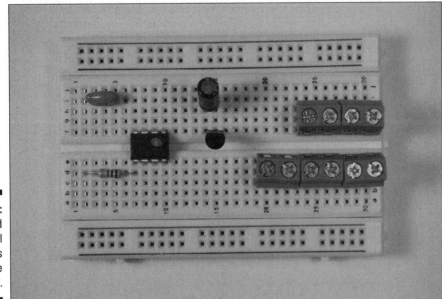

Figure 11-9:
Add
terminal
blocks
to the
breadboard.

3. **Insert wires to connect each component and terminal to the ground bus and insert a wire between the two ground buses to connect them, as shown in Figure 11-10.**

   On the resonator, the center of the three leads is the ground lead.

   Eight shorter wires connect components to ground bus; the long wire on the left connects the two ground buses.

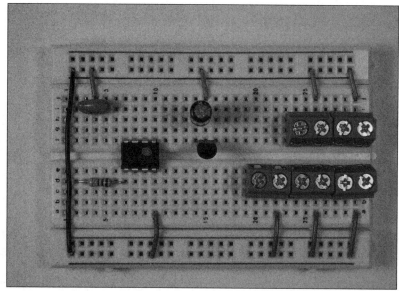

Figure 11-10:
Short wires connect components to ground bus; a long wire connects the two ground buses.

4. **Insert wires to connect the input pin of the voltage regulator and the terminal block for the battery to the + voltage bus, as shown in Figure 11-11.**

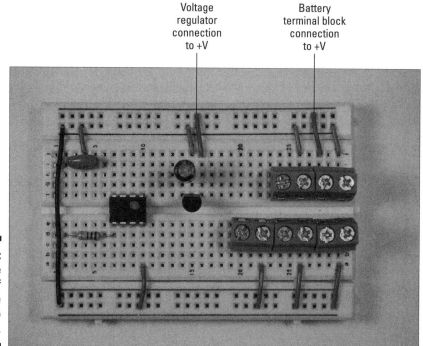

Voltage regulator connection to +V

Battery terminal block connection to +V

Figure 11-11:
Connect the input pin of the voltage regulator to the battery.

5. **Insert wires to connect the encoder IC, voltage regulator, and resonator, as shown in Figure 11-12.**

IC Pin 3 to the other outer resonator pin

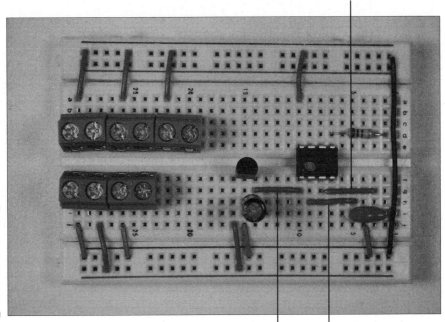

Figure 11-12:
Connect the IC, voltage regulator, and resonator.

Voltage regulator to IC Pin 1

IC Pin 2 to one outer resonator pin

6. **Insert wires to connect the terminal blocks to the IC1 encoder and resistor, as shown in Figure 11-13.**

Pin 7 of IC1 encoder to on/off terminal block   Resistor to LED terminal block

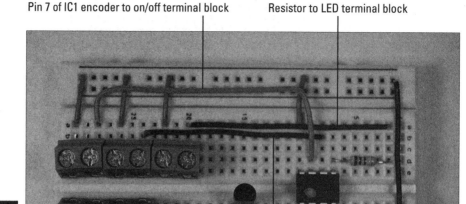

Figure 11-13:
Connect
terminal
blocks to
the IC1
encoder
and resistor.

Pin 4 of IC1 encoder to motor L terminal block   Pin 6 of IC1 encoder to motor R terminal block

7. **Drill holes in the transmitter box for the LED and switches.**

   Figure 11-14 shows holes drilled in the front of the box for the LED and in the top of the box for the switches. The size of drill bit that you use depends upon the diameter of the switches and LED socket you use. We used a ¼" drill bit to drill the hole for the LED and a ⁵⁄₁₆" drill bit to drill the holes for the switches.

Figure 11-14:
Drill holes
for the
LED and
switches.

8. **Place adhesive-backed Velcro in the box and on the battery pack, as shown in Figure 11-15.**

Figure 11-15:
Stick Velcro
to attach
the battery
pack.

9. **Place the three switches through the holes, as shown in Figure 11-16, and secure them to the box with the nuts included with the switches.**

Figure 11-16:
Transmitter
box with
switches
installed.

10. **Feed the LED socket through the hole in the front of the box. Slip the LED into the socket from the inside of the box, and then slip the bottom half of the socket over the leads and tighten to secure the LED.**

11. **Connect one black and one red 6" wire to the LED and solder them.**

The secured and soldered LED is shown in Figure 11-17.

The red wire must be connected to the longer of the LED leads. After the wires have cooled, wrap them in electrical tape to prevent the leads from shorting.

Heed all the safety precautions about soldering that we talk about in Chapter 2. For example, don't leave your soldering iron on and unattended. And for heaven sakes — don't drop it in your lap when it's hot!

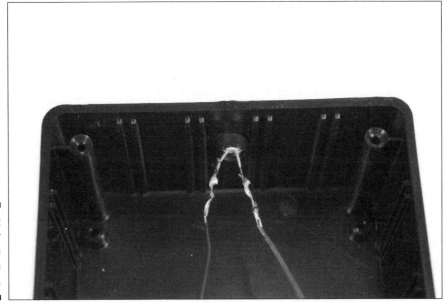

Figure 11-17:
Transmitter
box with
LED in
place.

12. **Connect 6" wires (any color is okay) to the switches and solder them, as shown in Figure 11-18.**

   To connect wires to the switches, feed the wire through the lug and twist to secure it.

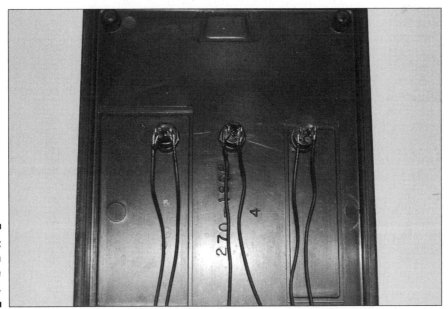

Figure 11-18:
Attach
wires to the
switches.

13. Secure the breadboard in the box with Velcro. Or, if the breadboard already has adhesive tape on the back like the one we used, just use the tape.

14. Attach the wires from the battery pack snap connector and the LED to the terminal blocks, as shown in Figure 11-19.

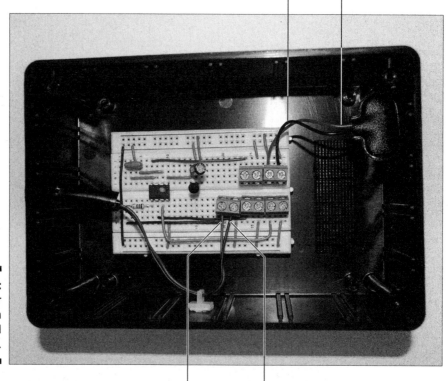

Black battery lead   Red battery lead

+V LED lead   Ground LED lead

Figure 11-19: Transmitter box with breadboard installed.

15. Attach the wires from the switches to the terminal blocks, as shown in Figure 11-20.

Leads to on/off switch   Leads to motor R switch

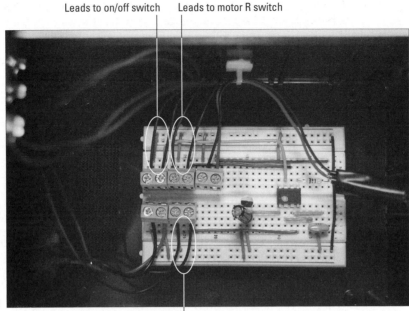

Figure 11-20:
Attaching
wires to
terminal
blocks.

Leads to motor L switch

16. **Place batteries in the battery holder, connect the snap, and secure the holder in the box with Velcro, as shown in Figure 11-21.**

Figure 11-21:
The inside
of the
completed
transmitter
box.

17. **Attach the cover to the box with the included screws.**

The completed box is shown in Figure 11-22.

Figure 11-22:
The
transmitter
all put
together.

Making the receiver circuit board

The receiver, quite logically, receives the signal sent by the transmitter. You will eventually affix the receiver circuit to the body of the go-kart, but don't put the kart before the horse (so to speak): Before you affix the receiver, you have to build it.

To build the receiver circuit, follow these steps:

1. **Insert the ICs on the breadboard, as shown in Figure 11-23.**

Be sure to leave room between each IC to place wires (as shown in the final figure in this section, the upcoming Figure 11-32). Also, leave room at the end of the board near the decoder to insert the IR detector and the resonator. Finally, leave room at the end of the board near the H-bridge to add terminal blocks.

H-bridge IC3   Inverter IC2   Decoder IC1

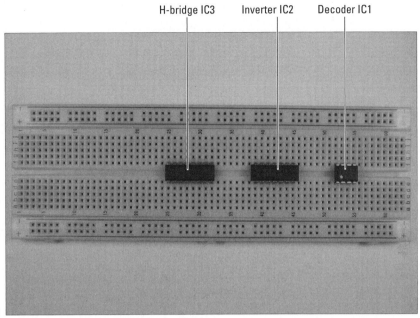

Figure 11-23:
Insert the
ICs on the
breadboard.

2. **Insert the voltage regulator, resonator, and IR detector on the bread-board, as shown in Figure 11-24.**

Voltage regulator   Resonator   IR detector

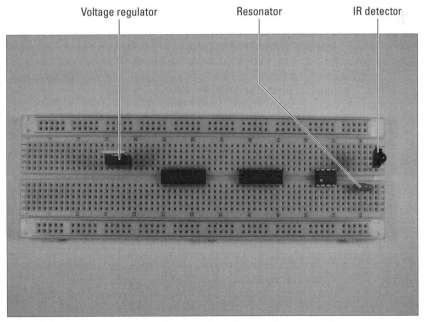

Figure 11-24:
Insert the
discrete
components.

Figure 11-25 shows how the circuit uses each pin of the voltage regulator and IR detector (referred to as *discrete components*).

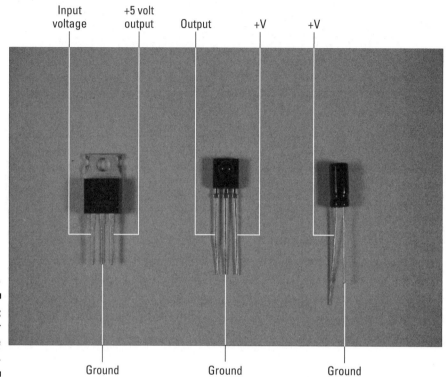

Input voltage   +5 volt output   Output   +V   +V

Ground   Ground   Ground

Figure 11-25: Pinouts for the discrete components.

3. **Insert four terminal blocks a few rows in from the end of the breadboard, as shown in Figure 11-26.**

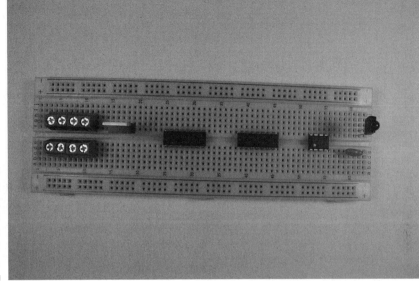

Figure 11-26:
Insert
terminal
blocks
on the
breadboard.

4. **Insert wires to connect each component to the + voltage bus and a wire to connect the two + voltage buses, as shown in Figure 11-27.**

   Note that you connect the +5 volt output of the voltage regulator to the + voltage bus because the voltage regulator takes the approximately +6 volts from the battery and changes it to +5 volts. It then supplies those +5 volts to the + voltage bus so that it can be used by the other components.

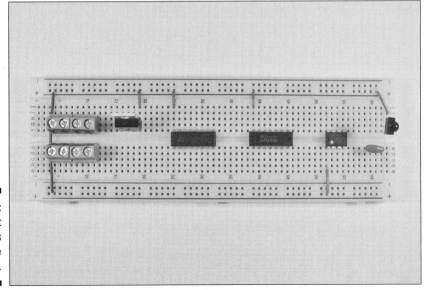

Figure 11-27:
Connect
components
to the
+V bus.

5. **Insert a set of one 10 microfarad and one 0.1 microfarad capacitor between the + voltage bus and the ground bus next to the V+ input for the decoder IC. Do the same for the inverter IC and the H-bridge IC, as shown in Figure 11-28.**

These capacitors filter out electrical noise introduced to the + voltage supply for each IC by the DC motors.

0.1 microfarad capacitors

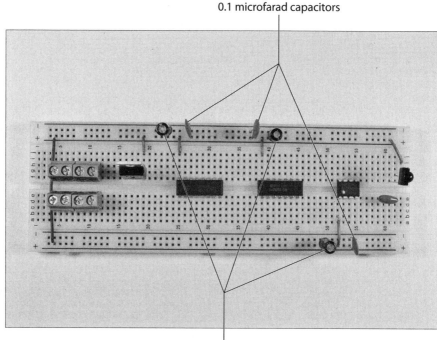

Figure 11-28:
Insert
capacitors
on the
breadboard.

10 microfarad capacitors

6. **Insert wires to connect each component and terminal to the ground bus (marked with a – on this breadboard) and insert a wire to connect the two ground buses, as shown in Figure 11-29.**

The resonator in the center of the three leads is the ground lead.

Wire connecting
2 ground buses

Voltage regulator
to ground bus

Decoder
to ground bus

Terminal block
to ground bus

H-bridge
to ground bus

IR detector
to ground bus

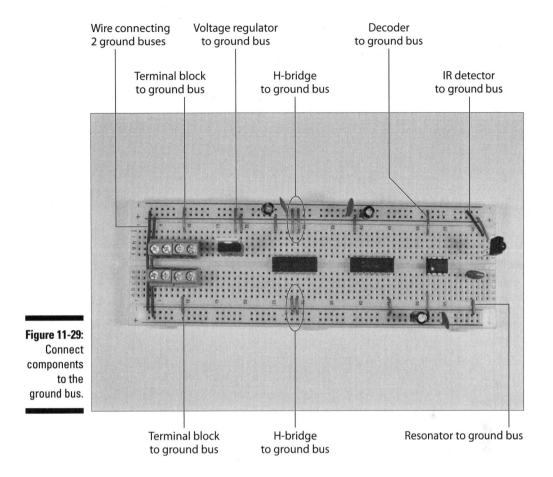

Figure 11-29:
Connect
components
to the
ground bus.

Terminal block
to ground bus

H-bridge
to ground bus

Resonator to ground bus

7. **Insert wires to connect the decoder IC, resonator, and IR detector, as shown in Figure 11-30.**

Pin 4 of the
decoder
to output pin
of the
IR detector

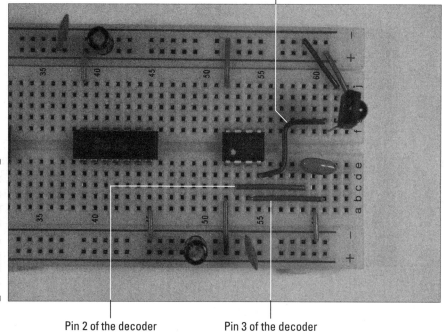

Figure 11-30:
Insert wires
between
pins on the
decoder,
resonator,
and IR
detector.

Pin 2 of the decoder
to the outer pin
of the resonator

Pin 3 of the decoder
to the other outer pin
of the resonator

8. **Insert wires to connect the decoder, inverter, and H-bridge, as shown in Figure 11-31.**

Pin 12 of the inverter to Pin 15 of the H-bridge

Pin 13 of the inverter to Pin 10 of the H-bridge

Pin 7 of the decoder to Pin 9 of the H-bridge

Pin 5 of the decoder to Pin 13 of the inverter

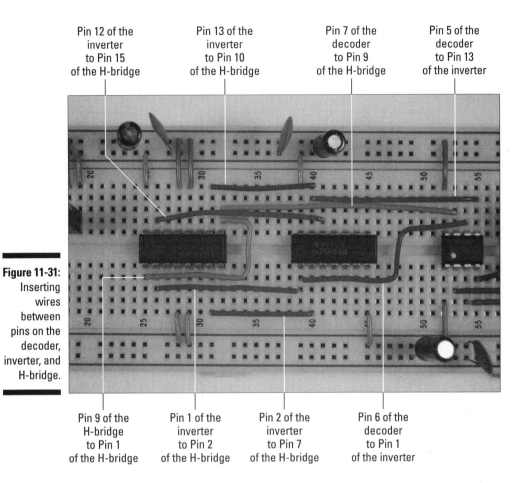

Figure 11-31:
Inserting wires between pins on the decoder, inverter, and H-bridge.

Pin 9 of the H-bridge to Pin 1 of the H-bridge

Pin 1 of the inverter to Pin 2 of the H-bridge

Pin 2 of the inverter to Pin 7 of the H-bridge

Pin 6 of the decoder to Pin 1 of the inverter

9. **Insert wires to connect the terminal blocks to the H-bridge and voltage regulator, as shown in Figure 11-32.**

Input pin
of voltage regulator
to +V pin
of battery
terminal block

Pin 14
of H-bridge
to other pin
of terminal block
for motor R

Pin 11
of H-bridge
to terminal
block
for motor R

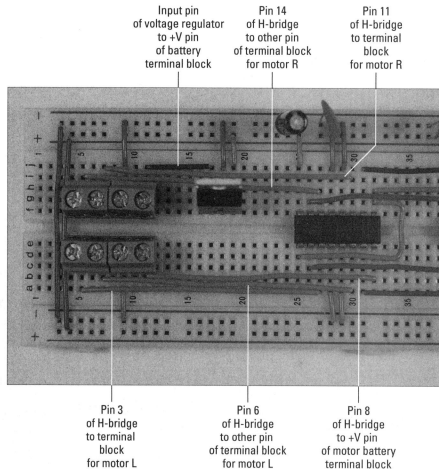

Figure 11-32:
Wiring the
terminal
blocks.

Pin 3
of H-bridge
to terminal
block
for motor L

Pin 6
of H-bridge
to other pin
of terminal block
for motor L

Pin 8
of H-bridge
to +V pin
of motor battery
terminal block

Building the go-kart

If you've followed this chapter to this point, all the little wires and doohick-eys in the transmitter and receiver circuit are in place to make your go-kart go. However, one thing is missing: the kart itself. This consists of a platform on which you affix

✔ The receiver

✔ The motors

✔ A switch to turn the power on and off

✔ Something to cover the chassis

Although the covering is optional, it makes the kart look cooler. More importantly, a covering prevents your kids or pets from poking a finger or paw where they shouldn't and pulling wires loose.

To build the kart itself, follow these steps:

1. **Place the breadboard, battery packs, caster, and power switch on the ½" PVC or plywood sheet, as shown in Figure 11-33.**

   This will help you to determine how large to make the base of the kart.

   We chose a ½" PVC sheet, 6½" wide and 9" long, which can hold all the components and still leave enough room to rest the top of the kart on the base. You can use a pencil to mark where to cut the ½" PVC sheet.

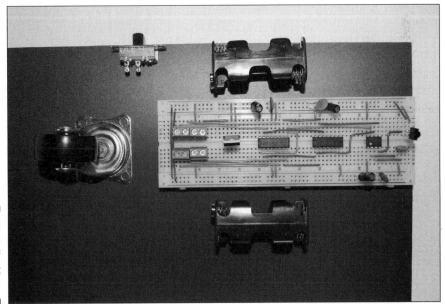

Figure 11-33: Determine the shape of the kart base.

2. **Mark a location between the breadboard and the bumper switch where you will drill a hole to feed the wires to the motors.**

3. **Use the base of the castor as a template to mark the four holes you will drill to mount it to the body.**

   Figure 11-34 shows the marked PVC sheet.

Figure 11-34:
Mark the base for cutting and drilling.

4. **Cut the sheet along the lines you drew (a hacksaw or any fine-toothed saw should do the job), and then use a ¼" bit to drill a hole where you marked the feed-through hole in Step 2.**

5. **Use a ⅛" bit to drill the holes used to attach the castor.**

   Use sandpaper or a file to smooth any rough edges. Figure 11-35 shows the base after it's been cut and drilled.

Figure 11-35:
The base for the kart, all sawed, drilled, and sanded.

6. **Attach Velcro to the battery packs, motors, and switch, as shown in Figure 11-36.**

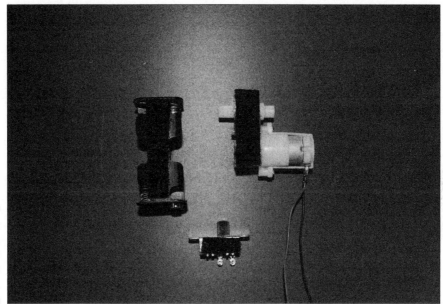

Figure 11-36:
Attach
Velcro to
the various
components.

7. **Solder 12" wires to the motors, as shown in Figure 11-37.**

By using different wire colors (we used red and black), it's easier to iden-tify which goes to each pin in the terminal blocks on the breadboard so you can control which direction the motors turn.

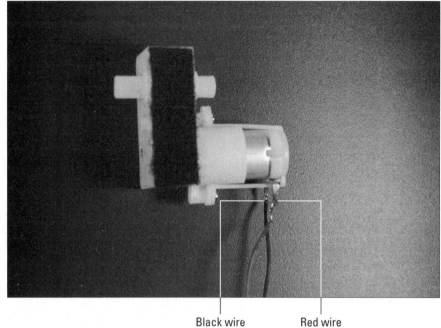

Figure 11-37:
Solder wires
to the
motors.

Black wire   Red wire

8. **Stick some Velcro to the base of the kart where the motors will be attached and then attach the motors; see Figure 11-38.**

Figure 11-38:
Attach the
motors and
castor.

9. Attach the castor, using ⅝" 6-32 screws and nuts, also shown in Figure 11-38.

10. Feed the wires through the ¼" hole and secure the wires with the wire clips (as shown in Figure 11-38); then slip the wheels on the motor shafts and secure the wheels with the screw provided.

11. Stick Velcro to the base of the kart and attach the battery packs, switches, and breadboard to the top of the base, as shown in Figure 11-39.

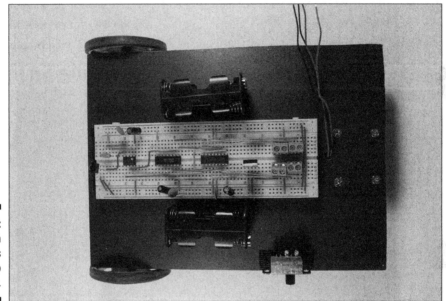

Figure 11-39:
Attach various parts to the kart.

12. Attach wires from the motors to the terminal blocks, as shown in Figure 11-40.

Red wire from motor R   Red wire from motor L

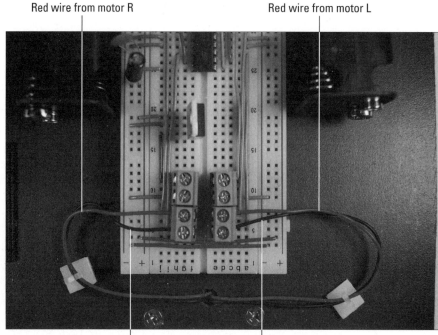

Figure 11-40:
Wire the
motors
onto the
breadboard.

Black wire from motor R   Black wire from motor L

13. **Attach wires from the battery packs to the terminal blocks and power switch, as shown in Figure 11-41, and solder the wires attached to the power switch lugs.**

Red wire
from battery pack

Red wire
from battery pack

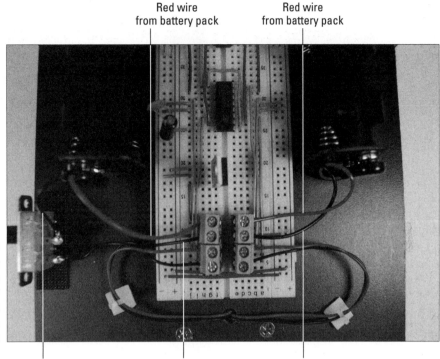

Figure 11-41:
Wire
the power
onto the
breadboard.

Black wire
from battery pack
to power switch

Black wire
from power switch

Black wire
from battery pack

14. With a mini hacksaw, cut openings in the back of the plastic kart top.

One opening will allow the IR detector to peek out and detect a signal; the other opening is for the power switch in the side.

Make sure you're using a top made of flexible plastic so it cuts easily. Wear safety glasses in case a piece flies off while you're cutting.

15. Rest the bubble on the kart, as shown in Figure 11-42.

Figure 11-42:
The final
product:
an infrared
go-kart!

Trying It Out

Here's the payoff to all your hard work: You finally get to see the little kart zoom around your living room. Remove any obstacles before you start!

Follow these steps to operate your go-kart:

1. **Slip in the batteries.**

2. **Turn on the power switch on the kart.**

3. **Point the transmitter at the kart and press and release the on/off button.**

4. **Watch it go!**

If nothing happens, here are a few things to check out:

- ✔ All the batteries are fresh, are tight in the battery pack, and face the right direction.

- ✔ See whether any wires or parts have come loose.

- ✔ Compare your circuit with the photos in this chapter to make sure you got all the connections right.

To turn left, press and release the motor L button; to turn right, press and release the motor R button. After the kart has turned far enough, press and release the same button again, and the kart will go forward. If you want the kart to go backward, just press and release motor button L and quickly press and release motor button R (or vise versa); ditto if the kart is moving backward and you want it to move forward.

If these buttons work opposite to the way you expect, swap the wires from the motors at the terminal blocks on the receiver.

Just like your car on a cold winter morning, the kart could take a few moments to warm up. Wait a few beats after you turn on both on/off switches, and then go for it!

If you plan on keeping the kart working long-term (rather than using the parts on another project), we suggest that you remove the Velcro from the motors and from the base (where you mounted the motors) and glue the motors to the kart base at the same spot. This will make the kart a little more stable. You can also glue the wheels onto the motor shaft to make your kart more permanent.

Taking It Further

Are you so wowed by the go-kart you want to add to it? Here are some ideas to explore:

- ✔ Other encoder/decoder ICs are available that have more than the three input/output pins used here. One option is the Holtek HT12A encoder and HT12D decoder; another is Reynolds Electronics IR-DX8 encoder/decoder. Building a kart by using one of these lets you add other things to the kart that you can control: for example, LED headlights or a buzzer to use just like a car horn.

- ✔ If you want to race go-karts with a friend, you can build multiple karts with one of the other encoder/decoder pairs just mentioned. These allow you to pick an address for each kart by tying different pins to ground or +V on each of the encoder/decoder pairs. However, you still have to be careful not to activate your transmitters at the same time or point your transmitter at your buddy's car lest you confuse the receivers. The only way around this is to use radio control instead of infrared control, and use different frequencies for each transmitter. See Chapter 18 for information about using radio control.

- ✔ Spiff up the go-kart base and top any way your artistic whim dictates: Paint it, add racing stripes, use a different shape, add decals, or glue fuzzy fur and cute little ears on it.

Part IV
Good Vibrations

In this part . . .

The chapters in this part all deal with things that vibrate in some fashion, from sound waves in the air to a furniture cushion. Intrigued? Here's what's covered:

Radio waves form the basis of the remote control unit that you use to manage a character we call *Sensitive Sam* (Chapter 13). Oscillating signals sent by a metal detector (Chapter 12) produce a corresponding oscillating signal in metal, helping you identify when you've hit gold (or more likely, iron). Finally, our Couch Pet-ato (Chapter 14) responds to the vibration that occurs when your cute kitty jumps up on your couch in your absence, producing a sound to scare off the little dear.

Chapter 12

A Handy-Dandy Metal Detector

*F*inding little bits of metal is a time-honored task, right up there with finding free music to download off the Internet. With a metal detector, you can scout out coins in your couch, nails in your walls, keys in somebody's pocket — face it, the applications are endless.

In this project, we show you how to build a small, handheld metal detector that can detect certain kinds of metal — especially ferrous (iron-containing) metals — even if that metal is under a half-inch or so of drywall, sand, or soil (but never water!).

The Big Picture: Project Overview

In this project, you use an IC that generates an AC signal that goes through a coil. Because metal objects conduct electricity, you can induce a current in those objects. When the coil in the metal detector comes near a metal object, the electromagnetic field in the coil induces currents in the metal object. The electromagnetic field generated by the metal changes the current in the coil. When the signal changes, IC turns on an LED, alerting you to the presence of metal.

You can see the finished metal detector in Figure 12-1.

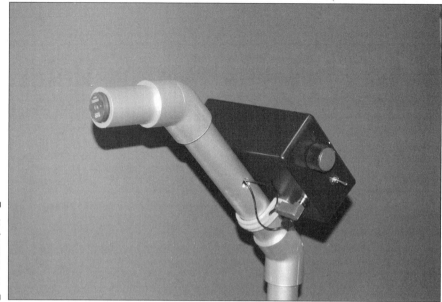

Here are the types of activities you'll be doing to build the metal detector. You will

1. Build a pretty simple electronic circuit containing a coil, a proximity detector IC, a transistor, a couple of resistors, and an LED.

2. Install the circuit in a box with batteries and an on/off switch.

3. Mount the box on a handle made from plastic pipe.

Scoping Out the Schematic

This project is really easy. You have only one breadboard to put together for this project. You can see the schematic for the board in Figure 12-2.

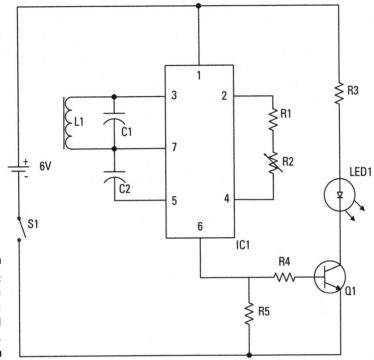

Here's a list of the schematic elements:

✔ **L1** is a coil (also called an *inductor*) wired in parallel to capacitor C1; the combination of this dynamic duo is a parallel *LC* (inductor/capacitor) circuit. When a signal that oscillates at several kHz passes through this circuit, the signal creates an electric field around the coil. When you bring the coil near a metallic object, that electric field induces an oscillating signal in the object. Turnabout is fair play, so when the oscillating signal has been induced in the metallic object, the signal in the object creates an electric field that induces current in the coil. This current changes the oscillating signal running through the LC parallel circuit.

✔ **IC1** is a TDA0161 proximity detector. This IC is designed to supply the oscillating signal that's sent through the LC parallel circuit. The IC also responds to any changes in the signal: the IC has an output of 1 milliamp (mA) or less if the coil is far from a metallic object and an output of 10 mA or higher if the coil is near a metallic object.

✔ **R1** is a resistor, and **R2** is a variable resistor. These resistors are used to calibrate the circuit in IC1 to the LC circuit. You calibrate the circuit by adjusting the value of the variable resistor when the coil is not near any metal objects.

- ✔ **R5** is a resistor connected between the output of IC1 and ground. When the output of IC1 is on, current flows through this resistor and provides a positive voltage to the base of Q1.

- ✔ **Q1** is a 2N3904 transistor that you connect to the output of IC1. When the output of IC1 is high, Q1 turns on and allows current to flow through LED1.

- ✔ **LED1** is the indicator light used to indicate that the device has detected metal in the vicinity.

- ✔ **R3** is a resistor that limits the amount of current flowing through LED1, which prevents it from burning out.

- ✔ **S1** is the on/off switch.

Building Alert: Construction Issues

You use PVC cement (a glue) to connect the PVC fittings that form the microphone handle. You can get this glue at any building supply store.

This glue creates a very strong joint. However, be sure to wear some form of work gloves when using this glue because it melts plastic — and you definitely don't want it on your hands. Also read the label for safety tips, such as using the glue in a well-ventilated area and what to do if some comes in contact with your skin.

Perusing the Parts List

Even though this project doesn't require many parts, you still have to go out and buy or assemble them. Several of the parts are shown in Figure 12-3. Here's what you need:

- ✔ **Two 1 kohm resistors (R1, R4)**
- ✔ **10 kohm potentiometer (R2)**
- ✔ **330 ohm resistor (R3)**

- 120 ohm resistor (R5)
- Two 0.0047 microfarad ceramic capacitors (C1, C3)
- One 2N3904 transistor (Q1)
- One T-1 ¾ LED (LED1)
- LED panel-mount socket, T-1 ¾
- TDA0161 proximity detector (IC1)
- Battery pack for 4 AA batteries

- 680 picofarad bobbin-type inductor (L1)

  We used C&D Technologies part #1468420C that we picked up at Mouser (www.mouser.com).

- SPST (single-pole, single-throw) toggle switch, used as the on/off switch
- 400-pin breadboard
- Four 2-pin terminal blocks
- Knob (for the potentiometer)
- Two phono jacks

- Two right-angle phono plugs

  We used right-angle plugs to avoid having a loop of wire coming out of the box. You can also use banana plugs and jacks.

- Enclosure

  We used a plastic box, RadioShack part #270-1806.

- An assortment of different lengths of prestripped, short 22 AWG wire
- Two PVC 45° joints with 1" slip fitting on both ends (one male and one female)
- PVC 1" end cap with 1" female slip fitting
- 1" diameter PVC pipe (referred to as *schedule 40 pipe*), 1' in length
- 1" clamp to attach the enclosure to the pipe
- Two ½" 8-32 panhead screws
- Two 8-32 nuts

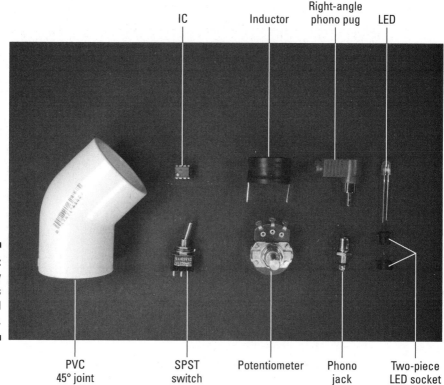

IC   Inductor   Right-angle phono pug   LED

Figure 12-3:
Key
components
of the metal
detector.

PVC
45° joint   SPST switch   Potentiometer   Phono jack   Two-piece LED socket

Taking Things Step by Step

Although the circuit is simple, there are a few steps you need to do to assemble it and then put the whole shooting match together, including building the handle and attaching the circuit to it. As usual, we start with the circuit.

Building a metal detector circuit

The circuit in this project controls sending a detector signal and processing the signal that comes back to light up the LED. Here are the steps involved:

1. **Place TDA0161 (IC1), 2N3904 (Q1), and four terminal blocks (TB) on the breadboard, as shown in Figure 12-4.**

   The transistor pinout is shown in Figure 12-5.

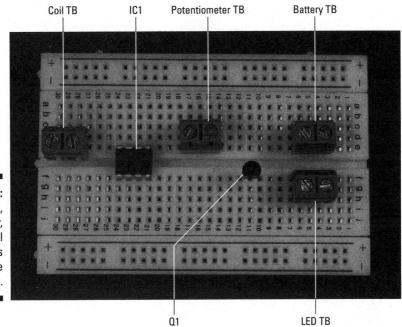

Figure 12-4:
Place the IC,
transistor,
and terminal
blocks
on the
breadboard.

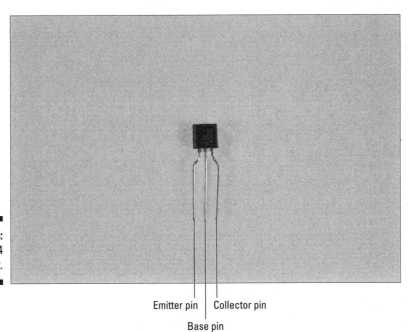

Figure 12-5:
The 2N3904
pinout.

2. **Insert wires to connect the battery terminal block and the transistor emitter pin to the ground bus. Then insert a wire between the two ground buses to connect them, as shown in Figure 12-6.**

   Two shorter wires connect components to ground bus; the long wire on the right connects the two ground buses.

Figure 12-6:
Connect
components
to ground
bus; then
connect the
two ground
buses.

3. **Insert wires to connect IC1 and the battery terminal block to the +V bus. Then insert a wire between the two +V buses to connect them, as shown in Figure 12-7.**

Pin 1 of IC1 to +V   Battery terminal block to +V

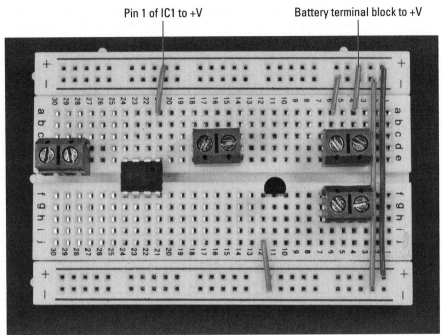

Figure 12-7:
Connect
components
to the
+V bus.

4. **Insert wires to connect the IC and discrete components, terminal block for the coil (L1), terminal block for the potentiometer (R2), and terminal block for the LED, as shown in Figure 12-8.**

L1 TB
to Pin 3 of IC1

R2 TB
to Pin 4 of IC1

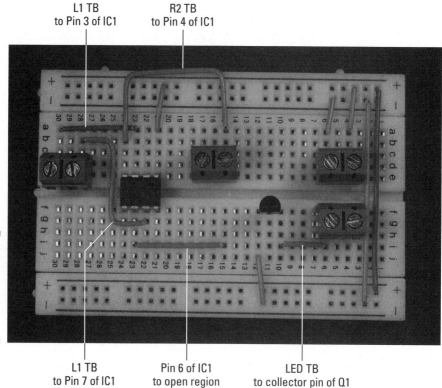

Figure 12-8:
Hook up the
ICs, terminal
blocks (TB),
and discrete
components.

L1 TB
to Pin 7 of IC1

Pin 6 of IC1
to open region

LED TB
to collector pin of Q1

5. **Insert two 0.0047 microfarad capacitors (C1 and C2), two 247 ohm resistors (R3 and R5), one 1 kohm resistor (R1), and one 120 ohm resistor (R5) on the breadboard, as shown in Figure 12-9.**

We discuss in Chapter 4 how to determine how short to clip the leads of many of these components to make them fit neatly on the breadboard. Make sure you wear your safety glasses when clipping leads!

C1 between pins
of L1 TB

R1 from Pin 2
of IC1 to R2 TB

R4 from base pin of Q1
to Pin 6 of IC1

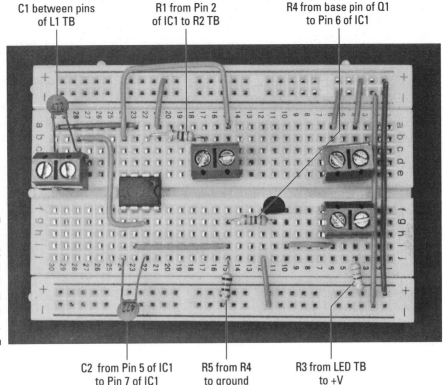

Figure 12-9:
Insert
resistors
and
capacitors
on the
breadboard.

C2 from Pin 5 of IC1
to Pin 7 of IC1

R5 from R4
to ground

R3 from LED TB
to +V

Building the box to house the circuit

The box that houses the circuit needs holes so that you can place the LED, the potentiometer dial you use to adjust resistance on the IC, and the on/off switch as well as various wire connections.

Follow these steps to get the metal detector circuit enclosure ready:

1. **Drill holes in the box where you will mount the LED, potentiometer, audio jacks, on/off switch, and clamp.**

   We put the on/off switch and the potentiometer on one side of the box, the LED on one end, and the audio jacks on the bottom. However, the placement is really up to you. Figure 12-10 shows where we placed these components.

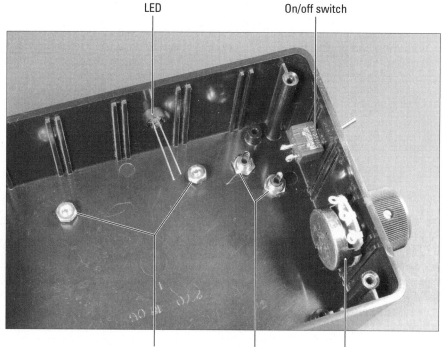

Figure 12-10:
Box with
on/off
switch,
phono jacks,
potentiome-
ter, and LED
in place.

LED   On/off switch

Screws and nuts securing the clamp   Phono jacks   Potentiometer

See Chapter 4 for more information about choosing drill bit sizes for particular components and other advice about how to customize a box for
your projects. And guess what? We always advise that you use safety
glasses when drilling.

2. **Slip the threads of the phono jacks through the drilled holes and
   secure with the nuts provided.**

3. **Slip the shaft of the on/off switch through the drilled hole and secure
   with the nut provided.**

4. **Slip the shaft of the potentiometer through the drilled hole and secure
   with the nut provided.**

5. **Slip the knob on the potentiometer shaft and secure with the set screw.**

6. **Feed the top half of the LED socket through the drilled hole from outside the box and insert the LED into the top half of the socket from
   inside the box.**

7. **Slip the bottom half of the socket over the leads and snap onto the top
   half of the socket to secure the LED.**

8. **Solder the black wire from the battery pack to one lug of the on/off
   switch and solder an 8" black wire to the remaining lug of the on/off
   switch, as shown in Figure 12-11.**

Figure 12-11:
Wires
soldered to
the on/off
switch,
potentio-
meter,
phone jacks,
and LED.

9. **Solder an 8" wire to the center potentiometer lug and another 8" wire to the left potentiometer lug, as shown in Figure 12-11.**

10. **Solder a red 8" wire to the long lead of the LED and a black 8" wire to the short lead of the LED, as shown in Figure 12-11.**

11. **Slip a 1" piece of heat shrink tubing over each solder joint and use a hair dryer to secure them in place.**

12. **Solder an 8" wire to the lug on each of the phono jacks, as shown in Figure 12-11.**

See Chapter 2 for advice about safe soldering if you're not very experienced in this art.

Putting it all together

After you have a box and a circuit, it's time to introduce them to each other. Follow these steps to do so:

1. **Attach Velcro to the breadboard and the box and secure the breadboard in the box.**

2. **Attach Velcro to the battery pack and the box and secure the battery pack in the box.**

3. Insert the wires from the LED, potentiometer, on/off switch, and battery pack to the terminal blocks on the breadboard, as shown in Figure 12-12.

Black wire from on/off switch

Wires from Potentiometer   Red wire from battery pack

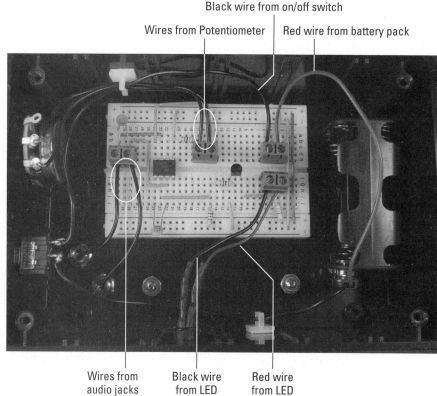

Figure 12-12:
Connect the LED, potentiometer, on/off switch, coil, and battery pack to the breadboard.

Wires from audio jacks   Black wire from LED   Red wire from LED

As you insert the wires, cut each of them to a sufficient length to reach the assigned terminal block and strip the insulation from the end of the wires.

4. Secure the wires with wire clips where needed.

Handling the handle

To easily wander around pointing the metal detector at suspected deposits of metal, you need a handle. Here are the steps to do so:

1. **Glue an 8", 1"-diameter PVC pipe into one end of a 45°-angle PVC pipe fitting, facing up. (See the upcoming Figure 12-13.)**

2. **Glue the other 45° PVC pipe fitting onto the other end of the PVC pipe, facing down. (See the upcoming Figure 12-13.)**

3. **Glue a 3", 1"-diameter PVC pipe into the open end of one of the 45° pipe fittings to form the coil end of the metal detector.**

4. **Glue a 6", 1" PVC pipe into the open end of the other 45° fitting.**

   This becomes the handle end of the metal detector.

5. **Glue the 1" PVC cap on the open end of the 6" PVC pipe.**

6. **Drill a ⅜" hole in the middle of the long section of PVC pipe on the side that will be to your left when you're holding the metal detector.**

   You use this hole to feed the wires from the microphone cartridge to the box containing your circuit. *Note:* If you're left handed, consider placing this hole on the side that will be to your right when you're holding the detector so that you hold the handle in your left hand and operate the switches with your right hand.

   Figure 12-13 shows the PVC pipe and fittings made into a handle for the metal detector.

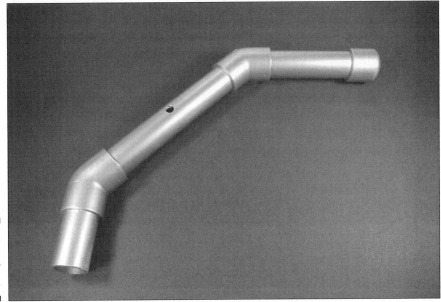

Figure 12-13:
Metal
detector
handle.

7. Solder 12" wires to each of the two inductor leads, as shown in Figure 12-14. Slip a 1" segment of heat shrink tubing over each solder joint and use a hair dryer to secure the tubing in place.

8. Twist together the free ends of the wires from the inductor and feed them through the PVC pipe from the coil end until the end of the wire strand reaches the ⅜" hole.

9. Form a hook shape with a piece of 20 or 22 gauge wire and pull the wires through the ⅜" hole.

10. Insert the inductor in the end of the 1" PVC pipe, as shown in Figure 12-15, and use some glue to secure the inductor in the pipe.

Figure 12-15:
Inductor in
the end of
the PVC
pipe.

11. **Cut the wires to allow three inches to extend from the ⅜" hole in the pipe and attach each wire to a right-angle phono plug, as shown in Figure 12-16.**

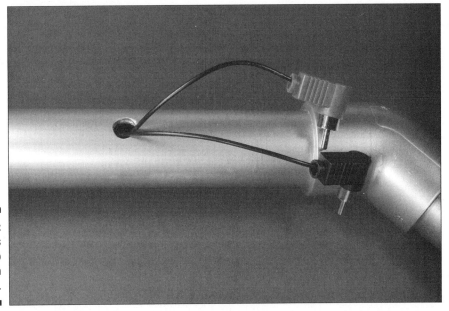

Figure 12-16:
Phono plugs
attached to
wires from
inductors.

You can use either a plug that requires soldering to the wire or one that uses a screw to secure the wire, as we have here.

12. **Press the clamp onto the 1" PVC pipe and attach the box to the clamp with the 8-32 screws and nuts.**

Figure 12-17 shows the box attached to the handle.

13. **Plug the right-angle phono plugs into the phono jacks, as shown in Figure 12-17.**

Figure 12-17:
Box attached to the handle and phono plugs in place.

The finished metal detector is shown in Figure 12-18.

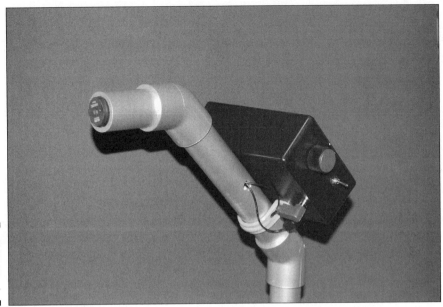

Figure 12-18:
The finished
metal
detector.

Trying It Out

You probably have a fortune in coins just waiting to be found at the bottom of your furniture cushions, so it's time you got this metal detector thing on the trail of all those nickels and dimes! Here's how to make this gadget work:

1. **Insert the batteries.**

2. **Secure the lid on the box with the screws provided and flip the on/off switch to On.**

3. **Holding the coil away from any metal, turn the potentiometer knob so that the LED is on, and then turn the potentiometer knob slightly in the other direction till the LED turns off.**

   This calibrates the IC so that it is triggered by small changes in the oscillating signal that runs through the coil.

4. **Try out your detector by holding it near different items containing metal.**

   We were able to detect coins and keys in pants pockets as well as various types of tools and nails at a distance of about $\frac{1}{2}$". You should be able to detect larger metal objects (such as a space shuttle) at a distance of about an inch.

If you don't get the results we got, here are some options to check out:

- ✔ Check that all the batteries are fresh, tightly inserted in the battery pack, and all face the right direction.
- ✔ Check that no wires or components have come loose.
- ✔ Compare your breadboard against the photos to make sure all the wires and components are connected correctly.

Taking It Further

Aren't metal detectors just the most addictive thing? (Well, maybe not, but they are kind of fun to play around with.) Here are some other things you can do to have fun with detectors:

- ✔ **Have your detector activate a buzzer instead of an LED by simply replacing the LED in the circuit with a buzzer.**
- ✔ **Make a more powerful detector that could find coins a few inches under the sand on the beach.**

  Check online for other metal detector circuits that have more oomph. www.thunting.com specializes in metal detectors you can use for hunting treasure, for example.

Chapter 13

Sensitive Sam Walks the Line

. . . .

. . . .

*O*kay, we have to admit this right upfront: This project is Earl's absolute favorite in this whole book. Sensitive Sam is a motorized cart. You stick black electrical tape on the floor to create a little path or track for Sam, and Sam uses his sensors to follow the tape around corners and in interesting loops you devise for hours of fun. He also has a little horn you can blow (to warn the cat that he's coming). What's not to love?

In this chapter, you discover how to give Sam the "eyes" he needs to sense where he's going and also how to build a radio remote control device to tell him what you want him to do. Although you'll find lots of little components and connections going on here, don't be intimidated; after you get going, we think you'll find it's a pretty fun project. (Earl did!)

The Big Picture: Project Overview

You're probably wondering what this Sam guy looks like and what he's capable of. Glad you asked. Here's the low down on Sam, who

✔ **Has three wheels:** This design makes the cart stable. If you use four wheels, you need to include a suspension mechanism to ensure that all wheels stay in contact with the floor at all times. We use one unpowered wheel in front and two independently powered wheels in back. This way, if the motor for one of the wheels in back is shut off, the cart turns in the direction of the motor that was shut off (left or right).

✔ **Sports two eyes:** These eyes help Sam figure out where to go. Sam's *eyes* — phototransistors pointed at the floor — sense infrared (IR) light that is sent out by IR LEDs and then reflected by the floor. By laying down a track of black electrical tape on the floor, you create an area that reflects less of the IR light.

We set up the circuit so that when Sam's eyes hover over reflective floor, the motors turn. If one eye is over the black tape or other nonreflective surface — for example, where a bend comes in the tape track — the motor connected to that eye shuts off, causing the cart to turn and follow the tape. When the eye is back over the reflective floor, the motor turns back on, and the cart goes in a straight line again.

✔ **Responds to a remote control:** This remote uses radio waves to send its commands to Sensitive Sam. (This is the same technology that a key chain remote control device uses to open the doors on a car.) You can set the switches on the remote control to tell Sam to turn on/off, slow down or speed up, or honk the horn. When you flip a switch and press the transmit button, that effect kicks in.

Sensitive Sam is shown in Figure 13-1 in all his glory.

Figure 13-1:
Our very
own
Sensitive
Sam and his
feline friend,
Willoughby.

Here are the types of activities you'll do in this project:

1. Put together the electronic circuit for the remote control transmitter and then fit the breadboard into a plastic box with switches.

2. Put together the electronic circuit that decodes the radio signal and controls the movement of Sam in response to his sensor eyes.

3. Mount the circuit onto a chassis along with DC motors, wheels, and a few switches.

In the end, you create a cart that follows a track all by itself and responds to your every command. It also has a cute little horn that you can toot.

Scoping Out the Schematic

You need to get your arms around two schematics to master this project. The first is the transmitter circuit that you use to send Sam his commands. The second is the receiver circuit that helps him understand what the heck you're saying!

Transmitting Sam's commands

You use the transmitter circuit to send signals to Sam to start, change speeds, and sound his horn. Here's an explanation of what's going on in this circuit:

- ✔ **VR1** is a voltage regulator that takes the 6 volts supplied by the battery pack and supplies a steady 5 volts instead. We added this because although the IC specs say it should work with 6 volts, it blew out the first time we tried using 6 volts. Better safe than sorry!

- ✔ The **transmitter module** sends out a radio signal at 433.9 MHz that's modulated with the code provided by the encoder.

- ✔ **IC1** is an encoder. The radio signal that the remote control sends out is modulated, depending upon how you have the switches set, by this encoding IC (see Figure 13-2). In this figure, the top line shows the code that tells Sam to speed up, and the bottom shows the code that tells Sam to slow down, based in the width of the fourth pulse from the right. The radio receiver in Sam sends this code to a decoding IC to turn Sam on/off, honk the horn, or change the speed.

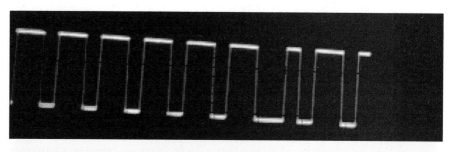

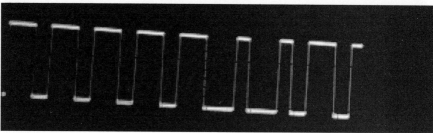

Figure 13-2:
The top
line speeds
Sam up;
the bottom
slows him
down.

✔ **Pin 14** is a transmit enable pin. When you press the normally open (NO) pushbutton switch between Pin 14 and ground, the encoder sends a signal to the transmitter module. This signal contains information that tells the decoding module whether the toggle switches (S1–S4) between Pins 10, 11, 12, and 13 and ground are open or closed.

✔ **R1** is a resistor that sets the frequency of an oscillator that's internal to the encoder. The signal from this oscillator is required to generate the encoded signal.

✔ **S5** is the on/off switch.

Figure 13-3 shows what's going on in the transmitter circuit.

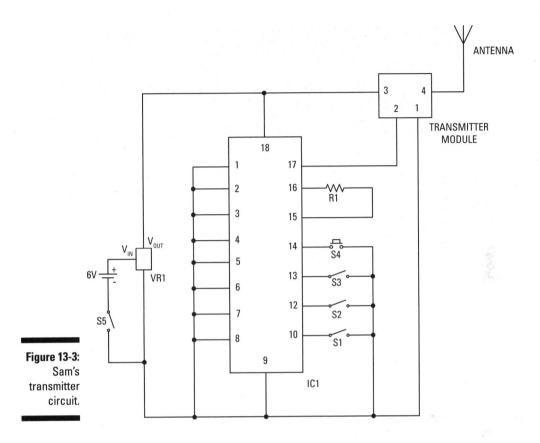

Figure 13-3:
Sam's
transmitter
circuit.

Helping Sam receive his commands

The receiver circuit that Sam uses to make sense of all your transmitted commands is shown in Figure 13-4. Here's how this one works:

✔ The **receiver module** separates the coded signal sent by the transmitter from the 433.9 MHz carrier wave. The output of the receiver module at Pin 2 is the decoded signal, as shown in Figure 13-2.

✔ **VR1** is a voltage regulator that takes the 6 volts supplied by the battery pack and supplies a steady 5 volts because the transmitter module has a maximum voltage of 5.5 volts. We provided a separate battery pack for this section of the circuit to provide a stable supply for the radio receiver. We found this setup gives much more consistent reception than setting up the receiver to share a battery pack with the rest of the circuit.

✔ **S1** is the on/off switch for the circuit.

✔ **IC1** decodes the signal sent by the transmitter. Pins 10, 11, 12, and 13 are outputs of the decoder that are at 0 volts if the corresponding toggle switch on the encoder is *closed* (connected to ground). The outputs are at about 5 volts if the corresponding toggle switch on the encoder is *open* (not connected to ground). These outputs stay at one voltage until they receive another signal to change.

✔ The **10 microfarad and 0.1 microfarad capacitors** (C1–C4 and C7–C12) placed between the +V and ground buses at the +V input of each of the ICs are used to filter electrical noise from the DC motors. That noise can prevent the ICs from getting the correct supply voltage.

✔ **IC2** is an LM555 timer that generates a square wave at Pin 3. This square wave alternates between 5 volts and 0 (zero) volts. The frequency of the square wave is determined by the values of **R2, R3,** and **C5**; we discuss this process in Chapter 9. When you tell Sensitive Sam to slow down, this square wave causes the voltage to the motor to switch on and off rapidly, with 6 volts to the motor half the time and 0 volts to the motor half the time. When you ask Sam to speed up, 6 volts are sent to the motor constantly. This is a simple form of *pulse width modulation,* which is commonly used to control the speed of DC motors.

✔ **C6** is a capacitor that reduces the occurrence of noise on Pin 5 of IC2, which could cause false triggering of the LM555. This might occur if Pin 5 were left unconnected.

✔ **Q1** turns the horn on and off. Q1 is a transistor that turns on when decoder Pin 12 is at 5 volts. Turning on Q1 allows a current to flow through the buzzer. To turn on the buzzer, you put the toggle switch connected to the encoder Pin 12 on the transmitter in the open position and push the transmit button. To turn off the buzzer, you put the toggle switch connected to encoder Pin 12 on the transmitter in the closed position and push the transmit button.

✔ **Q2** and **Relay 1** control Sam's speed. Q2 is a transistor that turns on when decoder Pin 10 is at 5 volts. Pin 4 and Pin 6 are normally connected (NC), and Pin 4 and Pin 8 are normally open (NO). Turning on Q2 allows current to flow through the coil in Relay 1 and connects Pin 4 to Pin 8. Pin 4 is the output of Relay 1, and Pin 8 is where you input the square wave from IC2 into Relay 1. When Q2 is off, no current flows through the coil in Relay 1, and Pin 4 is connected to Pin 6. This makes the output of Relay 1 approximately 5 volts. To ask Sam to go full speed ahead, you put the toggle switch connected to the encoder Pin 10 in the closed position and push the transmit button. To ask Sam to slow down, you put the toggle switch connected to the encoder Pin 10 in the open position and push the transmit button.

✔ Use **Q3** and **Relay 2** to tell Sam to start or stop. Q3 is a transistor that turns on when decoder Pin 13 is at 5 volts. Turning on Q3 by having the start/stop transmitter toggle switch closed allows current to flow through the coil in Relay 2 and connects Pin 4 (the output of Relay 2) to Pin 8. This tells Sam to run his engines. When the start/stop transmitter toggle switch is open and Q3 is off, no current flows through the coil in Relay 2, and Pin 4 is connected to Pin 6; this makes the output of Relay 2 zero (0) volts.

✔ **IC3** is an H-bridge motor controller. Although this controller is capable of controlling more functions than just going straight ahead (as you can read about in Chapter 11), all we need it to do here is supply the voltage to drive each motor forward. The battery pack attached to Pin 8 of IC3 supplies power for the motors. You connect the output of Relay 2 to the enable pins (1 and 9) of IC3. When 5 volts is provided by Relay 2 to the enable pins, IC3 supplies power to the motors. When 0 volts is connected to the enable pins, IC3 doesn't supply power, so Sam just sits there.

✔ The **left and right sensors** allow Sensitive Sam to take control of himself. When the track curves or Sam drifts so that one of the sensors is over the black electrical tape, power is cut to the motor on that side. This causes Sam to move away from the tape. When the sensor again hovers over a reflective floor, power is restored to that motor, and Sam straightens out.

On the schematic, the orange (O) and green (G) wires connect to an LED, with **R4** and **R6** limiting the current to protect the LED from damage. The blue (B) and white (W) wires connect to a phototransistor. When the sensor moves over a reflective surface, such as hardwood floor, the phototransistor is on, and the base of **Q5** or **Q4** is connected to ground. This turns off Q5 or Q4, which leaves the output of **Relay 3** or **Relay 4** connected to Pin 11, allowing the motor to run. When the sensor is over a nonreflective surface, such as black electrical tape, the phototransistor is off, and the base of Q5 or Q4 is connected to a positive voltage through **R7** or **R5**, turning on Q5 or Q4. This disconnects the output of the relay from Pin 11 and also shuts off the motor.

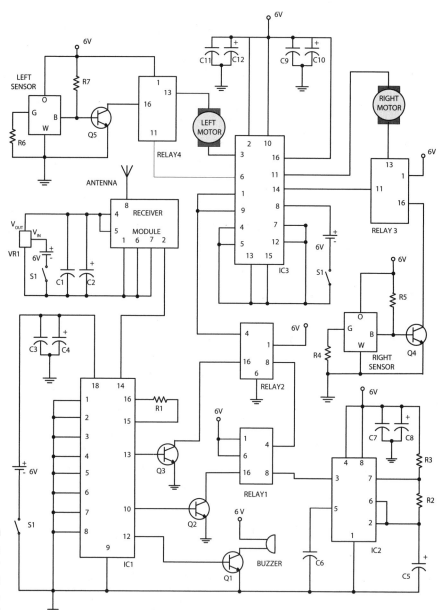

Figure 13-4:
The receiver
schematic
revealed.

Building Alert: Construction Issues

Sam is sensitive, and so are some of the issues you'll encounter when building him. For example, the motor lugs used in this project are made of thin metal and will break off if you put too much stress on them. By using stranded wire, rather than solid wire, you can minimize the stress on the lugs. Throughout the construction instructions that follow, we indicate when to use stranded wire.

Another construction issue to be aware of is the antenna. You solder antenna wire to one lead on both the transmitter and receiver modules. A 12" 20 gauge wire makes a dandy antenna; unlike 22 gauge wire, 20 gauge wire is stiff enough to stay upright. The only complication is in soldering the wire to the leads on the modules, so here are a few tips:

- Keep the soldering time to a few seconds to avoid damaging some of the solder joints in the module.
- Leave ¼" of lead below the solder joint to allow you to insert the adjacent leads fully into the breadboard.
- Soldering the 20 gauge solid wire directly to the lead is acceptable; however, soldering a few inches of stranded wire to the end of the antenna, covering that joint with heat shrink tubing, and soldering the stranded wire to the module lead prevents the leads from being twisted while you work with them.

Twisting can put too much stress on the lead and break it off, which is more likely to happen if you solder solid wire directly to the lead.

Perusing the Parts List

We broke down the parts shopping list into two . . . um . . . parts: a list for the transmitter circuit parts and a list for the receiver circuit and container parts.

Tallying up transmitter bits and pieces

The circuit that sends signals to Sam telling him what to do involves the following parts, several of which are shown in Figure 13-5:

- **LM7805 5 volt voltage regulator (VR1)**
- **Four SPST toggle switches (S1, S2, S3, S4)**

✔ **SPST normally open (NO) momentary push button switch (S4)**

✔ **Holtek HT12E encoder (IC1)**

✔ **1 megohm resistor (R1)**

✔ **TWS-434 RF transmitter module**

We bought this at Reynolds Electronics (www.rentron.com); Hobby Engineering (www.hobbyengineering.com) carries a similar module.

✔ **400-contact breadboard**

✔ **One 4 AA battery pack with snap connector**

✔ **Five 2-pin terminal blocks**

✔ **Plastic box**

We use Radio Shack part #270-1806.

✔ **An assortment of different lengths of prestripped short 22 AWG wire**

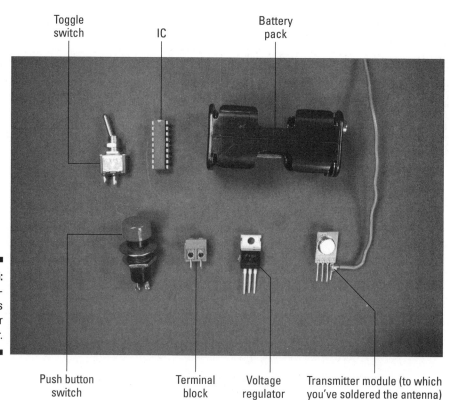

Toggle switch   IC   Battery pack

Figure 13-5:
Key components of your transmitter.

Push button switch   Terminal block   Voltage regulator   Transmitter module (to which you've soldered the antenna)

Running down receiver and container parts

The circuit that takes transmitted signals and explains to Sam what's expected of him involves the following parts, several of which are shown in Figure 13-6:

- ✔ **Holtek HT12D decoder (IC1)**

- ✔ **L293D H-bridge (IC3)**

- ✔ **LM555N-1 timer (IC2)**

- ✔ **Five 2N3904 transistors (Q1–Q5)**

- ✔ **6 volt buzzer**

- ✔ **RWS-434 RF receiver module**

  We bought this at Reynolds Electronics; Hobby Engineering carries a similar module.

- ✔ **Four 1 amp or greater solid state relays, DPDT** (double-pole, double-throw) **or SPDT** (single-pole, double-throw)

  We used the Shinmei RSB-5-S DPDT that we found at Jameco (www. jameco.com). A SPDT would also work, but we used the DPDT because it allows for more flexibility for which side of the relay we could run wires to. Make sure that the relay you buy has a pinout pattern that fits a breadboard; many of them do not.

- ✔ **Two DC gear motors GM2 each with a 2⅝" wheel or equivalent**

  We use these because the suppliers (Hobby Engineering (www.hobby engineering.com) or Solarbotics Ltd. (www.solarbotics.com) carry wheels made to fit them.

- ✔ **Two metal brackets used as motor mounts**

  We found 3" x ⅝" mending braces made by National Manufacturing Company at our local hardware store. These worked great.

- ✔ **One 1½" inch swiveling castor**

- ✔ **Six 0.1 microfarad ceramic capacitors (C1, C3, C6, C7, C9, C11)**

- ✔ **Six 10 microfarad electrolytic capacitors (C2, C4, C5, C8, C10, C12)**

- ✔ **51 kohm resistor (R1)**

- ✔ **Three 10 kohm resistors (R3, R5, R7)**

- ✔ **Two 150 ohm resistors (R4, R6)**

- ✔ **330 ohm resistor (R2)**

- ✔ **Two 830-contact breadboards**

- ✔ **Two Fairchild QRB1134 sensors**

- Three 4 AA battery packs with snap connectors
- Ten 2-pin terminal blocks
- Four 8-32 1½" panhead screws
- Four 8-32 nuts
- Four 6-32 ½" panhead screws
- Four 6-32 nuts
- Four 4-40 ¾" panhead screws
- Four 4-40 nuts
- Two wooden boxes
  - 2" wide x 5" tall x 1¼" deep
  - 5½" wide x 8½" long x 2½" deep

  We found one at a local craft supply store that was just the right size to hold the electronics for this project and a smaller wooden box to glue on the front of the bigger box to mount the sensors.

- An assortment of different lengths of prestripped, short 22 AWG wire

Mending brace   Buzzer   IC   On/off switch

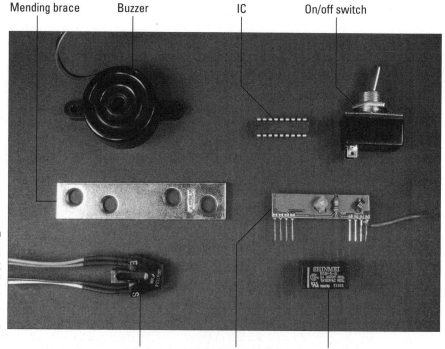

Figure 13-6:
Important
pieces of
the Sam
project.

Sensor   Receiver module   Relay
(with antenna soldered)

Taking Things Step by Step

The steps involved in making Sensitive Sam sensitive are

1. Making the transmitter circuit and fitting it into your remote control box
2. Making the receiver circuit that goes into Sam
3. Putting together the chassis that contains the receiver circuit (Sam's body)

Making the transmitter circuit and remote control box

The transmitter circuit fits into a remote control box and allows you to turn Sam on and off, speed him up or slow him down, and sound his horn. Here's what's involved in making this circuit:

1. **Place HT12E (IC1) and five terminal blocks on the breadboard, as shown in Figure 13-7.**

   The five terminal blocks shows in this figure will be used to connect two wires each to various components in the circuit. The wires from these terminal blocks will go to the battery pack, the on/off switch, the transmit switch, and the three toggle switches.

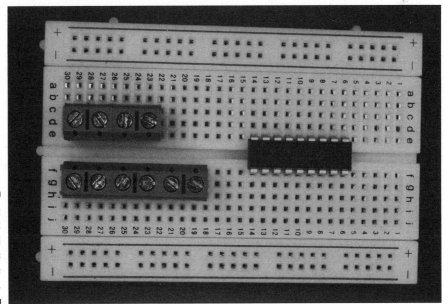

Figure 13-7:
Place the IC and terminal blocks on the breadboard.

2. **Solder an antenna wire to Pin 4 of the transmitter module, as shown in Figure 13-8.**

See the earlier "Building Alert: Construction Issues" section of this chapter for some tips on how to do this.

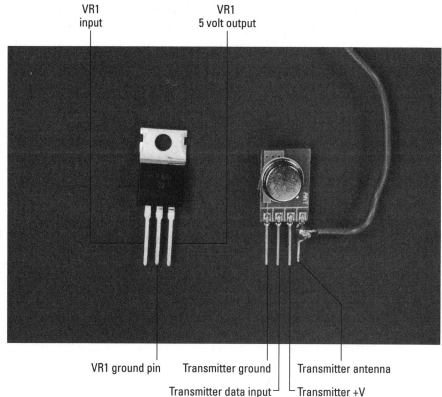

Figure 13-8:
Transmitter and voltage regulator (VR1) pinouts.

VR1 input

VR1 5 volt output

VR1 ground pin   Transmitter ground   Transmitter antenna

Transmitter data input   Transmitter +V

3. **Insert the voltage regulator (VR1), a 1 megohm resistor (R1), and the transmitter module on the breadboard, as shown in Figure 13-9.**

Voltage regulator

Figure 13-9:
Insert a
resistor,
voltage
regulator,
and the
transmitter
on the
breadboard.

R1 from Pin 16 of IC1
to Pin 17 of IC1

Transmitter

4. **Insert wires to connect the IC, voltage regulator, transmitter, and the terminal blocks to the ground bus. Then insert a wire between the two ground buses to connect them, as shown in Figure 13-10.**

Sixteen shorter wires connect components to ground bus; the long wire on the right connects the two ground buses.

Figure 13-10:
Make
ground bus
connections.

5. **Insert wires to connect the IC, voltage regulator, and transmitter to +V; then insert a wire between the two +V buses to connect them, as shown in Figure 13-11.**

Output pin of voltage regulator to +V

Figure 13-11:
Connect
components
to the
+V bus.

Pin 18 of IC1   Pin 3 of transmitter
to +V   to +V

6. **Insert wires to connect the IC, voltage regulator, transmitter, and terminal blocks, as shown in Figure 13-12.**

S1 TB to Pin 10 of IC1

Battery TB to VR1 input pin │ Pin 17 of IC1 to transmitter data input pin

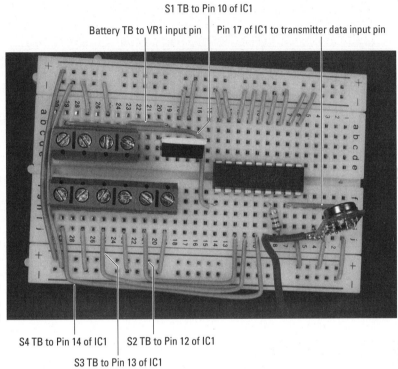

Figure 13-12:
Hook up the
ICs, terminal
blocks
(TBs),
transmitter,
and voltage
regulator.

S4 TB to Pin 14 of IC1 │ S2 TB to Pin 12 of IC1

S3 TB to Pin 13 of IC1

The next step is to drill all kinds of holes into which you can pop various components to create the remote control box. Follow these steps to do so:

1. **Drill holes in the box where you will mount the on/off switch, speed switch, horn switch, start/stop switch, transmit switch, and antenna, as shown in Figures 13-13 and 13-14.**

You can rearrange the switches. Just be careful not to put a switch where it will be in the way of the battery pack when you mount it inside the box.

On/off switch   Speed switch   Start/stop switch   Horn switch   Transmit pushbutton

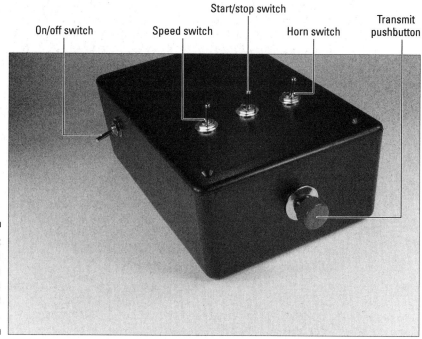

Figure 13-13:
Box with all the switches and buttons in place.

On the end of the right side of the box, you can see the hole used to feed antenna out of box.

Figure 13-14:
Use this hole to feed the antenna out of the box.

See Chapter 4 for more information about choosing drill bit sizes for particular components. In that chapter, we also offer advice about how to customize a box for your projects. Make sure you use safety glasses when drilling, and clamp the box to your worktable!

2. **Slip the shaft of the switches through the drilled holes and secure with the nuts provided.**

3. **Solder the black wire from the battery pack to one lug of the on/off switch and solder an 8" black wire to the remaining lug of the on/off switch, as shown in Figure 13-15.**

4. **Solder an 8" wire to each of the two lugs on the pushbutton transmit switch, as shown in Figure 13-15.**

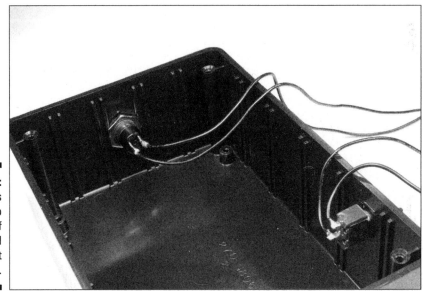

Figure 13-15:
Wires soldered to the on/off switch and transmit switch.

5. **Solder 6" wires to each of the two lugs on each of the three switches mounted on the cover, as shown in Figure 13-16.**

If you have 22 gauge stranded wire, consider using it for connecting the switches mounted on the cover. Stranded wire is more flexible, which makes getting the cover on the box easier. As we discuss in Chapter 4, solder the end of the stranded wire to gather all those loose strands.

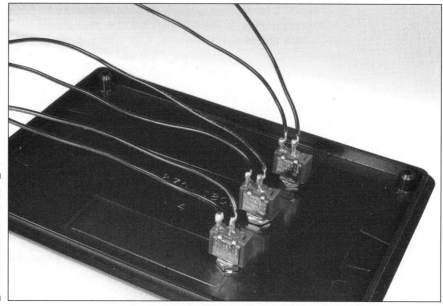

Figure 13-16:
Wires
soldered to
the speed,
horn, and
start/stop
switches.

Repeat after us: Heed all the safety precautions about soldering that we give you in Chapter 2. Use adequate ventilation when soldering to avoid inhaling fumes, and be sure to get a soldering iron with a stable stand so there's no danger of it falling off your work surface.

6. **Secure the breadboard and battery pack in the box with Velcro, as shown in Figure 13-17.**

   As you place the breadboard in the box, carefully feed the antenna wire out of the box through its hole and secure the antenna wire with a wire clip.

7. **Attach the wires from the battery pack, on/off, transmit, speed, horn, and start/stop switches to the terminal blocks, as shown in Figure 13-17.**

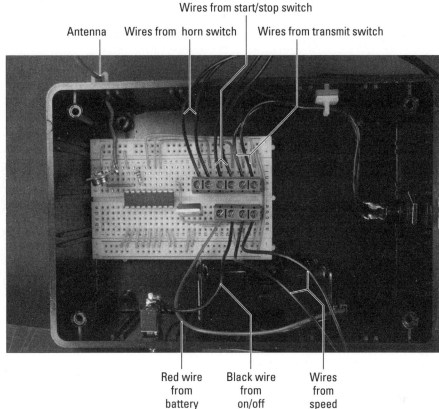

Antenna Wires from horn switch Wires from start/stop switch Wires from transmit switch

Figure 13-17:
Transmitter
circuit in
place.

Red wire
from
battery
pack

Black wire
from
on/off
switch

Wires
from
speed
switch

Making the receiver circuit

The receiver circuit will eventually reside in Sam's body, taking the signals that your remote control unit sends telling Sam how to behave. Here are the steps to build the receiver circuit:

1. **Take two 830-contact breadboards and join them by inserting the tabs into the slots along the sides to make one large breadboard.**

2. **Place the HT12D (IC1), the LM555 (IC2), the L293D (IC3), the receiver module, and ten terminal blocks on the breadboard, as shown in Figure 13-18.**

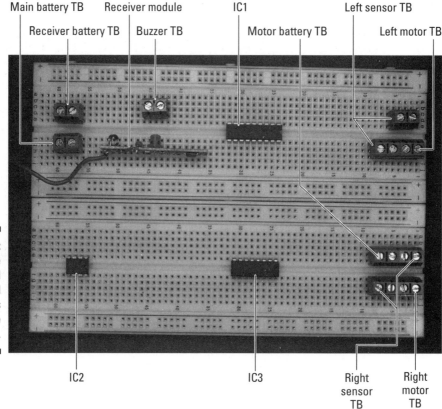

Main battery TB   Receiver module   IC1   Left sensor TB

Receiver battery TB   Buzzer TB   Motor battery TB   Left motor TB

Figure 13-18:
Place the
ICs and
terminal
blocks
on the
breadboard.

IC2   IC3   Right sensor TB   Right motor TB

3. **Place the four RB5-5-S relays (Relay 1–Relay 4) and five 2N3904 (Q1–Q5) transistors on the breadboard, as shown in Figure 13-19.**

   Insert each transistor lead into a separate breadboard row with the

   - Collector lead to the left side, as shown in Figure 13-19
   - Base lead in the center
   - Emitter lead to the right

   The collector pin of Q2 goes in the same breadboard row as Pin 16 of Relay 1. The collector pin of Q3 goes in the same breadboard row as Pin 16 of Relay 2. The pin designations of the 2N3904 transistors and the LM7805 voltage regulator are shown in Figure 13-20.

VR1   Q1   Relay 4   Q5

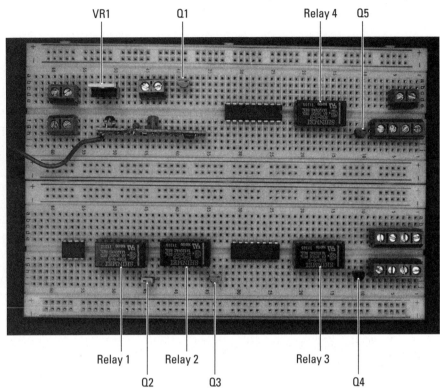

Figure 13-19:
Place the
RB5-5-S
relays,
LM7805
voltage
regulator,
and 2N3904
transistors
on the
breadboard.

Relay 1   Relay 2   Relay 3

Q2   Q3   Q4

Pin 11
(right NC)

Pin 16
(coil ground)

Pin 13
(right
common)

Pin 9
(right NO)

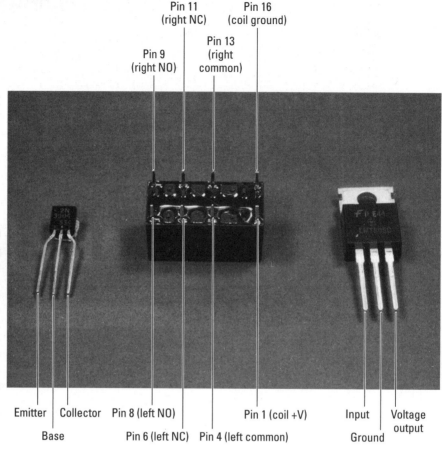

Figure 13-20:
The 2N3904
transistor,
LM7805
voltage
regulator,
and RB5-5-S
relays
pinout.

Emitter | Collector

Base

Pin 8 (left NO)

Pin 6 (left NC) Pin 4 (left common)

Pin 1 (coil +V)

Input

Ground

Voltage
output

4. **Insert wires to connect the ICs, voltage regulator, transistors, relays, and terminal blocks to the ground bus. Then insert wires between the ground buses to connect them, as shown in Figure 13-21.**

The ground buses are designated by a negative (–) sign on this breadboard.

Thirty-one shorter wires connect components to ground bus; the two long and one short wires on the left connect the ground buses.

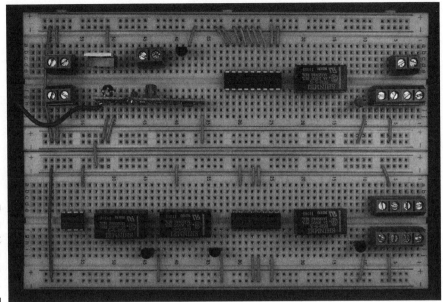

Figure 13-21:
Connect
components
to ground
buses.

5. **Insert wires to connect the ICs, the terminal blocks, relays, and the voltage regulator to the +V bus. Then insert wires between the +V buses to connect them, as shown in Figure 13-22.**

The +V buses are designated by a + sign on this breadboard.

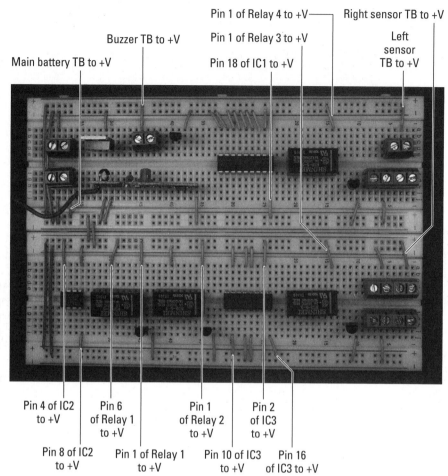

Pin 1 of Relay 4 to +V —— Right sensor TB to +V

Pin 1 of Relay 3 to +V   Left
                              sensor
Buzzer TB to +V   Pin 18 of IC1 to +V   TB to +V

Main battery TB to +V

Figure 13-22:
Connect
components
to the
+V bus.

Pin 4 of IC2   Pin 6   Pin 1   Pin 2
   to +V   of Relay 1   of Relay 2   of IC3
                    to +V   to +V   to +V

   Pin 8 of IC2   Pin 1 of Relay 1   Pin 10 of IC3   Pin 16
      to +V   to +V   to +V   of IC3 to +V

6. Insert wires to connect the ICs, terminal blocks, relays, and discrete components, as shown in Figures 13-23, 13-24, and 13-25.

The receiver module pinout is shown in Figure 13-26.

Receiver battery TB
to input pin of VR1

Buzzer TB
to collector pin of Q1

Pin 16 of Relay 4
to collector pin of Q5

Left sensor TB
to base of Q5

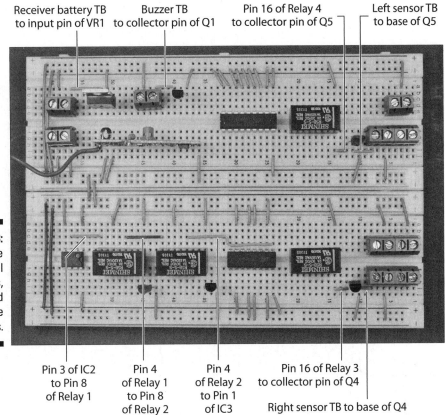

Figure 13-23:
Hook up the
IC, terminal
blocks,
relays, and
discrete
components.

Pin 3 of IC2
to Pin 8
of Relay 1

Pin 4
of Relay 1
to Pin 8
of Relay 2

Pin 4
of Relay 2
to Pin 1
of IC3

Pin 16 of Relay 3
to collector pin of Q4

Right sensor TB to base of Q4

Pin 5 of receiver module to Pin 4 of receiver module

Voltage output pin of VR1 to Pin 5 of receiver module

Pin 2 of receiver module to Pin 14 of IC1

Pin 13 of Relay 4 to left motor TB

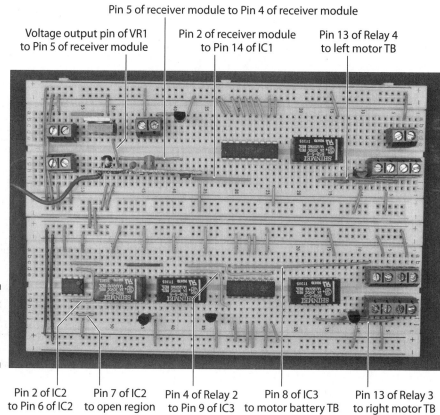

Figure 13-24:
Hooking up
yet more
wires.

Pin 2 of IC2 to Pin 6 of IC2

Pin 7 of IC2 to open region

Pin 4 of Relay 2 to Pin 9 of IC3

Pin 8 of IC3 to motor battery TB

Pin 13 of Relay 3 to right motor TB

Pin 10 of IC1 to base of Q2

Pin 6 of IC3 to Pin 11 of Relay 4

Pin 12 of IC1
to base of Q1

Pin 13 of IC1
to base of Q3

Pin 2 of IC3
to left motor TB

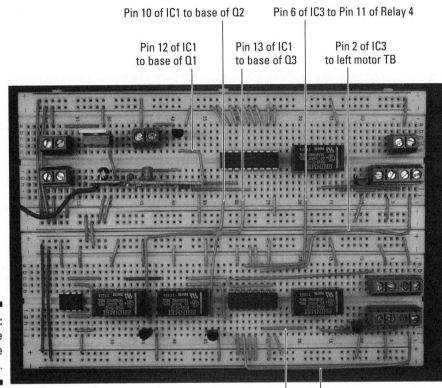

Figure 13-25:
Place the
rest of the
wires.

Pin 14 of IC3 to Pin 11 of Relay 3   Pin 11 of IC3 to right motor TB

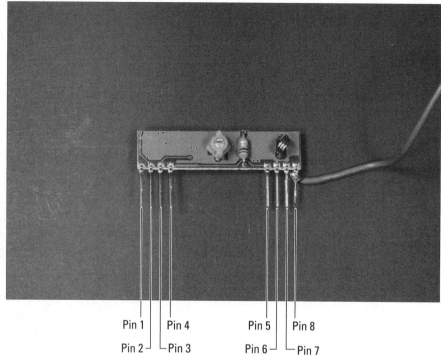

Pin 1 | Pin 4   Pin 5 | Pin 8
Pin 2 ⌐ ⌐Pin 3   Pin 6 ⌐ ⌐Pin 7

Figure 13-26:
Receiver
module
pinout.

7. Insert the following, as shown in Figure 13-27:

- Six 0.1 microfarad ceramic capacitors (C1, C3, C6, C7, C9, C11)
- Six 10 microfarad electrolytic capacitors (C2, C4, C5, C8, C10, C12)
- 51 kohm resistor (R1)
- Three 10 kohm resistors (R3, R5, R7)
- Two 150 ohm resistors (R4, R6)
- 330 ohm resistor (R2)

Use both the schematic and the photo to help you place each component. For example, the schematic shows that the + side of C5 is connected to Pin 2 of IC2 and the other side of C2 is connected to ground, so insert the long lead of C2 in the same row as Pin 2 of IC2 and the short lead in the ground bus.

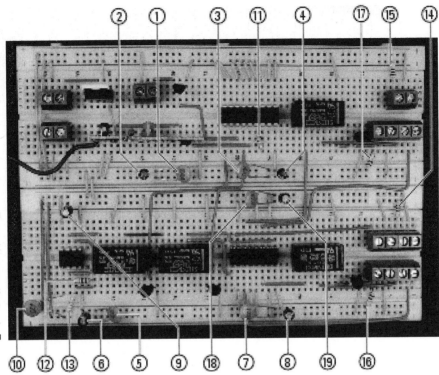

Figure 13-27:
Insert
capacitors
and
resistors
onto the
breadboard.

Use the following key for the callouts in Figure 13-27.

1. C1 from +V bus to ground bus
2. C2 from +V bus to ground bus
3. C3 from +V bus to ground bus
4. C4 from +V bus to ground bus
5. C7 from +V bus to ground bus
6. C8 from +V bus to ground bus
7. C9 from +V bus to ground bus
8. C10 from +V bus to ground bus
9. C5 from Pin 2 of IC2 to ground bus
10. C6 from Pin 5 of IC2 to ground bus

11. R1 from Pin 15 of IC1 to Pin 16 of IC1

12. R2 from Pin 6 of IC2 to Pin 7 of IC2

13. R3 from Pin 7 of IC2 to +V

14. R4 from right sensor TB to ground

15. R6 from left sensor TB to ground

16. R5 from right sensor TB to +V

17. R7 from left sensor TB to +V

18. C11 from +V bus to ground bus

19. C12 from +V bus to ground bus

Building Sensitive Sam's chassis

If you're a sensitive guy like Sam, how your body looks is very important to you. Here are the steps to build Sam a serviceable little chassis to hold the receiver circuit:

 1. **Solder 12" stranded wires to the motor lugs, as shown in Figure 13-28.**

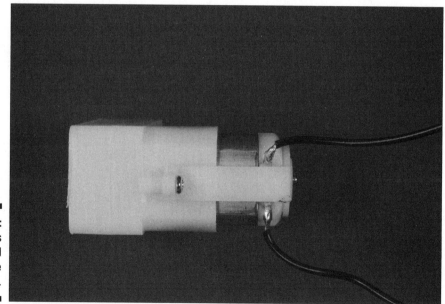

Figure 13-28:
Wires
soldered
to the
motor lugs.

2. **Drill holes for the 6-32 screws used to mount the castor, as shown in Figure 13-29.**

   Use the base of the castor to guide you when marking the four holes you will drill in the chassis box to mount the castor.

Holes for feeding wires   Screws for castor   Screws for motor mounts

Figure 13-29:
The inside of the box after holes have been drilled and the motors and castor are mounted.

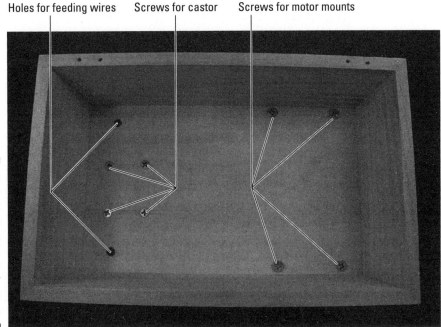

3. **Mark a location about an inch to each side of the castor and drill holes with a 1/4" bit to feed the wires to the motors and sensors, as shown in Figure 13-29.**

4. **Using a mending brace to mark the four holes you will drill in the box to mount the motors, drill holes for 8-32 screws, as shown in Figure 13-29.**

5. **Attach the castor to the box with four 6-32 screws and nuts, as shown in Figure 13-30.**

6. **Attach the motors to the box with two 8-32 screws and nuts and one mending brace for each motor, as shown in Figures 13-30 and 13-31.**

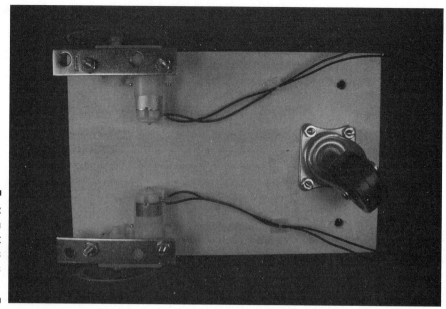

Figure 13-30:
The bottom
of the cart
with castors
and motors
mounted.

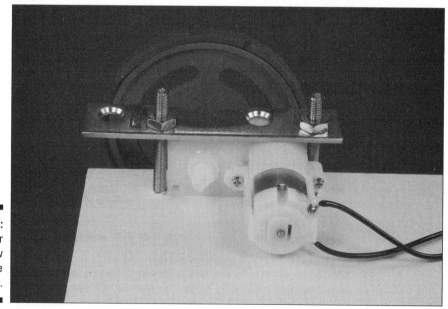

Figure 13-31:
A closer
look at how
to mount the
motors.

7. Drill a hole for 4-40 screws in each side of the small box to mount the sensors, as shown in Figure 13-32.

Figure 13-32:
Keeping the sensors just off the floor allows Sam to spot the tape.

8. **Mount the sensors in the small box with one 4-40 screw and nut per sensor.**

9. **Feed the wires for the sensors and the motors through the holes you drilled to each side of the castor.**

10. **Glue the small box to the cart, as shown in Figures 13-32 and 13-33.**

    The face of the sensors should be a quarter-inch above the floor.

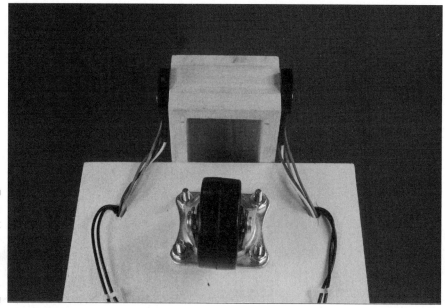

Figure 13-33:
A look at
how we
attached
the sensors
from below.

11. **Drill two holes for 6-32 screws to mount the buzzer, as shown in Figure 13-34.**

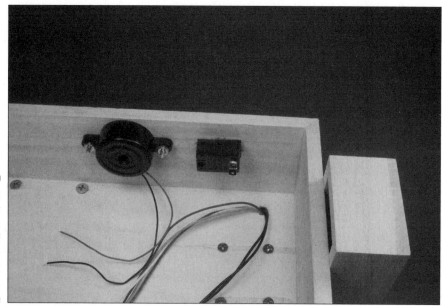

Figure 13-34:
The buzzer
and the
on/off
switch
mounted
on Sam.

12. **Mount the buzzer by using two 6-32 screws and nuts.**

13. **Drill a hole in the box where you will insert the on/off switch, as shown in Figure 13-34.**

14. **Slip the threaded shaft of the on/off switch through the hole you drilled and secure it with the nut provided.**

If the wall of the box is too thick to allow the threads on the switch to reach the nut, use a small chisel to remove enough wood from the inside wall of the box so that the nut can engage the threads.

15. **Solder the black wires from the three battery snaps to one lug of the on/off switch and solder three 12" black wires to the other lug of the on/off switch.**

Figure 13-35 shows the switch after soldering.

Figure 13-35:
Wires
soldered
to on/off
switch.

16. **Attach Velcro to the breadboard and the box and secure the breadboard in the box.**

17. **Attach Velcro to the battery packs and the box and secure the battery packs in the box.**

18. **Insert the wires from the sensors, motors, battery packs, buzzer, and the on/off switch to the terminal blocks on the breadboard, as shown in Figures 13-36 and 13-37.**

As you insert the wires, cut each of them to the length you need for them to reach the corresponding terminal block. Also, strip the insulation from the end of the wire.

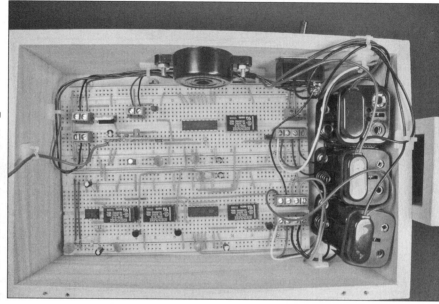

Figure 13-36:
Connect the sensors, motors, battery packs, buzzer, and on/off switch to the breadboard.

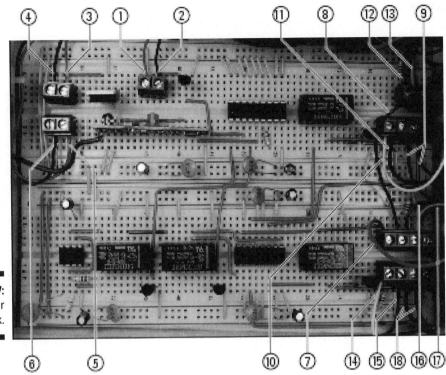

Figure 13-37:
A closer
look.

Use the following key for the callouts in Figure 13-37.

1. Red wire from buzzer

2. Black wire from buzzer

3. Red wire from receiver battery pack

4. Black wire from receiver battery pack

5. Red wire from main battery pack

6. Black wire from main battery pack

7. Red wire from motor battery pack

8. Black wire from motor battery pack

9. Wires from left motor

10. White wire from left sensor

11. Blue wire from left sensor

12. Green wire from left sensor

13. Orange wire from left sensor

14. White wire from right sensor

15. Blue wire from right sensor

16. Green wire from right sensor

17. Orange wire from right sensor

18. Wires from right motor

19. Secure the wires with wire clips where needed.

Sensitive Sam is shown roaring around our living room floor in all his glory in Figure 13-38.

Figure 13-38:
Sam chugs around the track till you tell him to stop.

Trying It Out

Now that you have a remote control unit and Sam's body all assembled, take him out for a spin. Follow these steps to play with your new sensitive buddy:

1. **Place black electrician's tape on a reflective floor (hardwood or linoleum, for example).**

   You don't have to create a straight line; you can use several pieces to design a circular or oval track.

2. **Place Sam on the track with the sensors on either side of the tape.**

3. **Put batteries in Sam and the remote control flip the on/off switches on both to On.**

4. **Flip the start/stop switch to start and press the transmit button on the remote control to get Sam moving.**

5. **Flip either the speed or horn switch on the remote control and then press the transmit button to activate either effect.**

6. **To stop Sam, flip the start/stop switch to stop and press the transmit button.**

If nothing happens, here are a few things to check out:

- All the batteries are fresh, are tight in the battery pack, and face the right direction.
- See whether any wires or parts have come loose.
- Compare your circuit with the photos in this chapter to make sure you got all the connections right.

If Sam gets going but doesn't follow the track as you expect, you can adjust the sensors by loosening the screws and sliding the sensors up or down.

If Sam stalls, try these steps:

1. **Put the start/stop switch in the start position.**

2. **Push the transmit button once more.**

Taking It Further

We're sure you can see why this neat little guy is Earl's favorite project. You can create huge tracks and have him follow around the room. He confuses recalcitrant cats (refer to Figure 13-1), and you can put notes in his cart and send them to someone else on the other side of the room.

When you're ready to take Sam further, try these suggestions:

- Build a Sensitive Samantha using a different radio frequency module for the remote control to give Sam a girlfriend he can race with around the track.

- Add lights by using the fourth pin on the encoder/decoder to control them.

- Read up on other radio control project ideas at sites such as www.renton.com.

- If you build Sam's chassis to be strong enough, you can put him to work carrying things around your house — a can of soda, the TV remote, or whatever you want to send off to the couch potato lounging in your living room.

Chapter 14

Couch Pet-ato

. . . .

In This Chapter

▶ Looking over the schematic

▶ Laying down the parts list

▶ Breadboarding the Couch Pet-ato circuit

▶ Installing the breadboard and components in the case

▶ Sounding off!

. . . .

You arrive home from a fast-paced game of racquetball, eager to grab a bag of chips and a cold can of soda, and then just veg out on your couch. As you walk in the front door, though, what do you find but your 30-pound cat, Reggie, spread out on the couch. Adding insult to injury, he's purring loudly — and shedding to beat the band.

Want to be saved from this annoying scenario forever after? That's what this chapter is all about: using a vibration/tilt switch sensor in a gadget that will sound off if Reggie (or Fluffy or Rover) so much as lays his paws on your furniture.

The Big Picture: Project Overview

When you complete this project, you'll have what we like to call a *Couch Pet-ato*. This device is a box rigged with a vibration sensor and recording/playback setup. You can record your voice, shouting something like, "Get the &!@!#@ off the couch!" or a shrill noise of your choosing. When your pet jumps onto the couch, the vibration sets off the playback.

You can see the finished Couch Pet-ato in Figure 14-1.

Figure 14-1:
The final
product:
a Couch
Pet-ato.

So what, exactly, will you be doing in this chapter? The project involves

1. Putting together the electronic circuit
2. Fitting the circuit into a wooden box
3. Attaching switches, a microphone, and a speaker

Scoping Out the Schematic

You have but one breadboard to pull together for this project. Take a look at the schematic for the board in Figure 14-2.

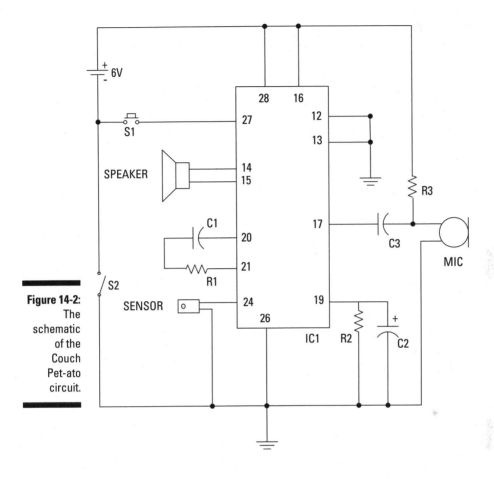

Here's the rundown of the schematic elements:

- The **sensor** — one of the key components of this circuit — is used to detect Fluffy jumping on the couch. This vibration/tilt switch sensor detects motion or vibrations when the switch is mounted with the body of the sensor horizontal to the bottom of the box. When the sensor detects motion, it closes a switch, just like how a toggle switch works.

- **IC1** is the other key component of this circuit. This is a chip that you can use to record a sound or voice message and play it back. We connected the sensor between Pin 24 of IC1 and ground. When the sensor detects motion and its contacts close, Pin 24 is connected to ground, which triggers the playback.

- **S1** is a normally open (NO) pushbutton switch that when depressed, connects Pin 27 of IC1 to ground. This causes the IC to record sounds or words that you speak into the microphone. Recording stops when you release the S1 pushbutton.

- ✔ **R3** is a resistor that connects the microphone to the +V, supplying the 4.5 volts that the microphone needs to function.

- ✔ **C3** is a capacitor that removes the DC voltage from the AC signal that's flowing from the microphone to Pin 17 of IC1.

- ✔ The **speaker** is connected to Pins 14 and 15 of IC1. The speaker is used to play messages that you recorded on IC1 when the sensor connects Pin 24 to ground.

- ✔ **S2** is the on/off switch between the negative terminal of the battery pack and the ground bus of the circuit board.

- ✔ **R1** and **C1** filter out electrical noise.

- ✔ **R2** and **C2** connect the automatic gain control circuit inside IC1 to ground. The values of R2 and C2 determine how fast the automatic gain control responds to changes in volume when you're recording a message.

Building Alert: Construction Issues

Because the Couch Pet-ato is likely to sit in your living room and we know that you like a neat-looking home, we placed all the works in a wooden box to make it look tidy.

We went strolling down the aisle of a local craft store to find our wooden box, but you might also find one in an office supply store — designed to hold index cards, for example. Just make sure that it's big enough to hold the breadboard.

Because the walls of this box are thicker than the plastic boxes often used for electronic gadgets, the threads on switches might not be long enough to secure them to the side of the box. If this happens to you, just drill the hole so that it's large enough to slip only the shaft of the switch through; then use glue to secure the switch.

The size of drill bit that you use to drill holes for the switches, speaker, and microphone depends upon the diameter of the components you're using. We used a ½" drill bit to drill the hole for the switches and a ⅜" bit to drill a hole for the microphone. For the speaker, we used a ⁵⁄₃₂" bit to drill the holes to which we attached screws for the speaker flange, and then we used a coping saw to cut the center hole for the speaker.

Perusing the Parts List

Here's your shopping list for building your Couch Pet-ato. This project involves the following parts, several of which are shown in Figure 14-3:

✔ **Signal Systems 3004 tilt/motion sensor**

Many vibration (also called *tilt* or *motion*) sensors are available; we used cost and pin type as our criteria. We found this sensor at Jameco (www. jameco.com), and Mouser (www.mouser.com) carries a similar low-cost sensor. We chose this one because it doesn't use a surface-mount package, which is harder to work with.

✔ **16 ohm, 0.2 watt speaker**

✔ **Electret microphone part #EM-99**

We found this one at Jameco. You can use other electret microphones; see Chapter 3 for the criteria to help you choose one. If you use another model of microphone, you might have to adjust R3 to get the supply voltage to the correct level.

✔ **Winbond Electronics ISD1110 voice record/playback chip (IC1)**

✔ **0.01 microfarad capacitor (C1)**

✔ **0.1 microfarad capacitor (C3)**

✔ **4.7 microfarad electrolytic capacitor (C2)**

✔ **Four ⅝" 6-32 screws**

✔ **Four 6-32 nuts**

✔ **Three wire clips**

✔ **5.1 kohm resistor (R1)**

✔ **470 kohm resistor (R2)**

✔ **2.2 kohm resistor (R3)**

✔ **NO (normally open) momentary pushbutton switch (S1)**

✔ **Breadboard (830 contacts)**

✔ **SPST toggle switch (S2)**

✔ **Four pack of AA batteries with snap connector**

✔ **Four 2-pin terminal blocks**

✔ **Enclosure**

We use a wooden box 8" x 5¼" x 3¼", with latches and clasp.

✔ **Velcro**

✔ **An assortment of different lengths of prestripped short 22 AWG wire**

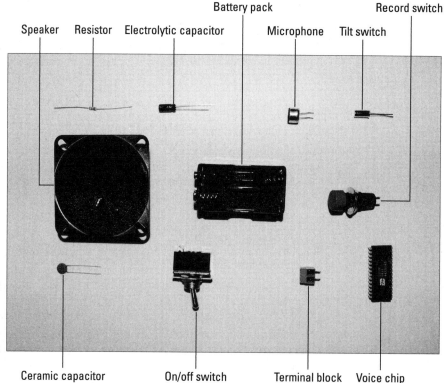

Battery pack
Record switch
Speaker   Resistor   Electrolytic capacitor   Microphone   Tilt switch

Figure 14-3:
Key
components.

Ceramic capacitor   On/off switch   Terminal block   Voice chip

Taking Things Step by Step

Creating your Couch Pet-ato involves wiring the circuit; installing the speaker,
microphone, battery, switches, and circuit board in the box; and then connecting the whole shebang.

Start at the beginning — by wiring the circuit. Follow these steps to build your
Couch Pet-ato circuit (but be sure not to let Rover know what you're up to).

1. **Place the voice chip IC and four terminal blocks on the breadboard,
   as shown in Figure 14-4.**

   The four terminal blocks shown in this figure will be used to connect
   two wires each to various components in the circuit. The wires from
   these four terminal blocks go to the battery pack, on/off switch, record
   switch, and microphone, respectively.

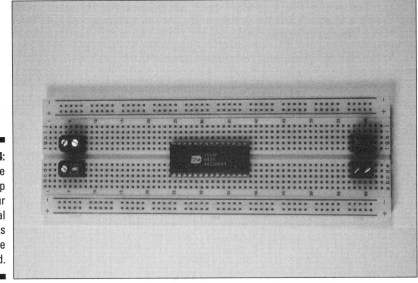

Figure 14-4:
Place the
voice chip
IC and four
terminal
blocks
on the
breadboard.

2. **Insert wires to connect each component and terminal to the ground bus and insert a wire between the two ground buses to connect them, as shown in Figure 14-5.**

In this figure, six shorter wires connect components to the ground bus (marked with a – on this breadboard); the long wire on the left connects the two ground buses.

Figure 14-5:
Connect
components
to the
ground bus.

3. Insert wires to connect each component and terminal to the +V bus and insert a wire between the two +V buses to connect them, as shown in Figure 14-6.

In this figure, four wires have been added:

- Three shorter wires connect components to +V bus (marked with a + on this breadboard).

- The long wire on the left connects the two +V buses.

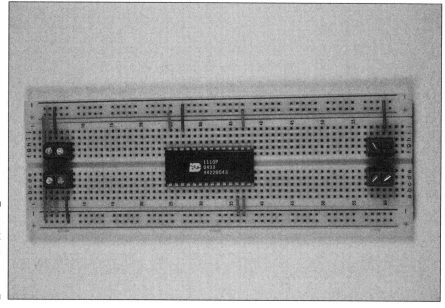

Figure 14-6:
Connect
components
to the
+V bus.

4. Insert wires to connect the voice chip IC to the terminal blocks and to the open regions of the breadboard where discrete components will be inserted, as shown in Figure 14-7.

open region of breadboard to terminal block for microphone

IC Pin 17 to open region of breadboard

IC pin 27 to terminal block for record switch

IC Pin 21 to open region of breadboard

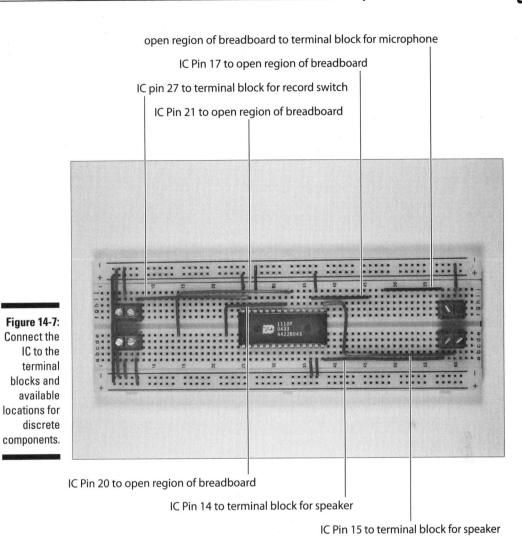

Figure 14-7:
Connect the
IC to the
terminal
blocks and
available
locations for
discrete
components.

IC Pin 20 to open region of breadboard

IC Pin 14 to terminal block for speaker

IC Pin 15 to terminal block for speaker

5. **Insert discrete components and the tilt sensor in the breadboard, as shown in Figure 14-8.**

Resistor R3 between capacitor C3 and +V

Resistor R2 between IC Pin 19 and ground

Capacitor C2 between IC Pin 19 and ground

Tilt sensor between IC Pin 24 and ground

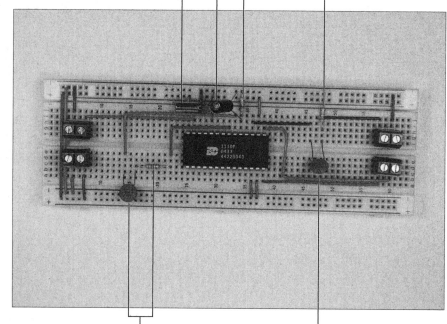

Figure 14-8:
Insert
discrete
components
on the
breadboard.

Capacitor C3 between wire connected to microphone
terminal block and wire connected to IC Pin 17

Capacitor C1 and resistor R1 in series between
wires connected to IC Pin 20 and IC Pin 21

6. Drill holes in the Couch Pet-ato box for inserting the switches, micro-phone, and speaker.

Figure 14-9 shows holes drilled in the front of the box for the switches and microphone and the side of the box for the speaker, but the placement is really up to you.

Whenever you drill holes or do any procedure where bits can fly, such as clipping wires, wear safety glasses to protect your eyes.

7. **Attach the speaker, switches, and microphone to the box, as shown in Figure 14-10.**

   We used ⅝" 6-32 screws and nuts to attach the speaker to the box. By creating a very tight hole for the microphone, we could press-fit it into the hole. (*Press-fitting* means that you don't need glue or another method to keep something in place.) The record switch was also press-fitted into a hole although we did use just a drop of glue to secure it. The on/off switch was attached with the supplied nut.

On/Off switch

Microphone

Figure 14-10:
Attach the
speaker,
switches,
and
microphone
to the box.

Speaker

Record switch

8. **Connect 6" wires (any color will work just fine) to the speaker and the record switch and solder them, as shown in Figure 14-11.**

Black wire to microphone

Wires to speaker   Wires to on/off switch

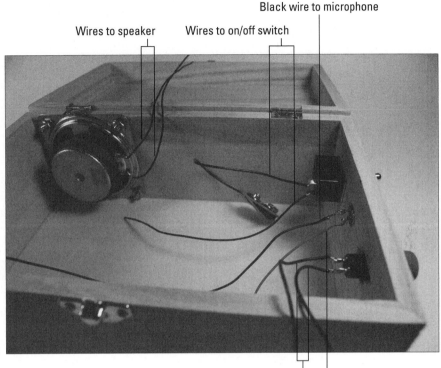

Figure 14-11:
Couch
Pet-ato box
with wires
soldered to
switches,
microphone,
and the
speaker.

Wires to record switch

Red wire to microphone

9. **Connect the black wire from the battery pack snap and another 6"
    black wire to the on/off switch and solder them together, as shown in
    Figure 14-11.**

10. **Connect a 12" red wire and a 12" black wire to the microphone and
    solder them, as shown in Figure 14-11.**

    Figure 14-12 shows which lead of the microphone you should solder the
    red and black wires to.

11. **After the wires cool, wrap the solder joints with electrical tape to pre-
    vent them from touching each other.**

Lead that red wire is attached to

Lead that black wire is attached to

Figure 14-12:
Identifying
the micro-
phone leads
to which
you will
solder
wires.

Don't get sloppy now! Heed all the safety precautions about soldering
that we provide in Chapter 2. For example, don't leave your soldering
iron on if you're not around. And for heaven sakes, don't drop it in your
lap when it's hot!

12. **Attach Velcro to the battery pack, breadboard, and box. Attach the
battery pack and breadboard to the box, as shown in Figure 14-13.**

   Note that we used three wire ties in the box to secure the wires and keep
   them out of the way.

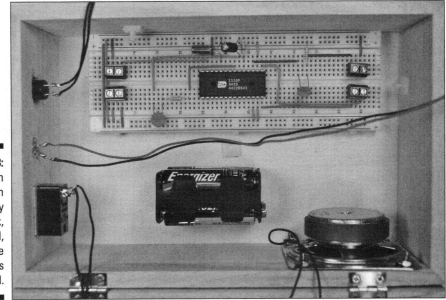

Figure 14-13:
Couch
Pet-ato with
battery
pack,
breadboard,
and wire
ties
attached.

13. **Attach the wires from the battery pack snap connector, on/off switch, record switch, microphone, and speaker to the terminal blocks, as shown in Figure 14-14.**

    When attaching the wires to the terminal blocks, cut the wires to the length you need and strip the ends.

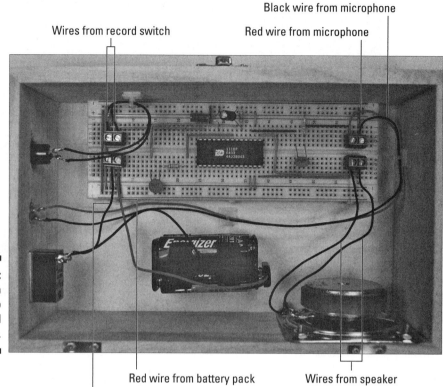

Wires from record switch

Black wire from microphone

Red wire from microphone

Figure 14-14:
Attach
wires to
terminal
blocks.

Red wire from battery pack

Wires from speaker

Wire from on/off switch

14. **Close and latch the lid.**

That's it! The completed box is shown in Figure 14-15.

Figure 14-15:
The Couch
Pet-ato,
all put
together.

Trying It Out

Okay, it's time for Fluffy or Rover to meet his or her newest adversary in the fight over couch real estate.

Follow these steps to operate your Couch Pet-ato:

1. **Slip in the batteries.**

2. **Flip on the on/off switch.**

3. **Press the recording button and hold it down while you speak your message or create a pet-annoying noise into the microphone.**

4. **Place the Couch Pet-ato on your furniture.**

5. **Drop your animal on the furniture (gently) and watch the fur fly!**

   Be sure that the Couch Pet-ato box is securely closed and latched. You don't want Fluffy to get into the works and chew up the wires or a resistor and choke.

If nothing happens, here is the usual list of a few things to check out.

- ✔ Check that all the batteries are fresh, are tight in the battery pack, and face the right direction.

- ✔ Check whether any wires or parts have come lose.

- ✔ Compare your circuit with the photos in this chapter to make sure that you got all the connections right.

If the phrase or sound that you recorded doesn't scare your pet, experiment with different phrases or louder noises. Cats hate hissing type sounds, and dogs respond better to verbal commands. Good luck!

Taking It Further

Our Couch Pet-ato (patent pending) is so cool that you might want to try some variations:

- ✔ If the sound isn't loud enough, add an amplifier between IC1 and the speaker. You can find out how that's done in an application note at the manufacturer's Web site at

  ```
  www.winbond-usa.com/products/isd_products/chipcorder/
            applicationbriefs/apbr06.pdf
  ```

  We did something similar in the circuit for Murmuring Merlin in Chapter 7. By adding the amplifier, the sound should be sufficiently dynamic to impress all but the most hard-of-hearing pets.

- ✔ Feel free to spiff up the Couch Pet-ato box by applying potato (or other) decals, painting it, or sticking glitter all over it.

- ✔ Try other uses for the Couch Pet-ato:

  - • Place it on top of your refrigerator and record a message like, "Remember your diet!" When somebody (who shouldn't) opens the refrigerator door headed for the leftover cheesecake, the brand-new *Pot Belly-tato* takes over.

  - • If you're a teen who fears that your parents might invade your bedroom space, add a strap to the Couch Pet-ato and hang it on your doorknob. When somebody opens your door without your permission, you can give him a piece of your mind. The verbal or other warning that you use in this *Back Off-tato* is entirely up to you!

Part V
The Part of Tens

The 5th Wave   By Rich Tennant

In this part . . .

Every *For Dummies* book has gotta have a few Part of Tens chapters! The chapters in this part use the ever-popular top-ten list format to give you a quick rundown of some useful information. Chapter 15 lists some of the better parts suppliers. Chapter 16 offers an assortment of great sources of information about electronics, from magazines to electronics clubs. And finally, Chapter 17 runs down ten great Web sites dedicated to electronics.

Chapter 15

Ten Great Parts Suppliers

*W*hen you enter the world of electronics projects, you begin a long-term relationship with electronics suppliers — the companies from which you buy resistors, transistors, sound chips, breadboards, and more.

All suppliers are not made equal. Some give better service, some offer better prices, and still others stock unique or hard-to-find items. In this chapter, we give some advice about what to look for in a supplier, listing ten or so suppliers we think are worth looking into.

It's probably worth ordering catalogs from all the suppliers listed here. The catalogs are free, and you can order them online. They come in handy for browsing because they organize things into categories that can help you find what you need. If you buy anything from a supplier's catalog, that firm usually sends you updates automatically. (And if nothing else, the hefty catalogs make great doorstops.)

When Is a Supplier Right for You?

If you're like us, sometimes you go to a large discount store for bargain bulk groceries; other times, you're off to a corner store because it's convenient or to a gourmet shop for a special (but expensive) treat. So how do you choose what electronics supplier to go to? What makes one electronics supplier better than another? Just like with food shopping, that might depend on your need of the moment.

Ordering online: If you like to order online, look for handy Web site features, such as being able to save your order and add to or modify it during multiple visits to the site. Many suppliers' Web sites offer this feature, but most offer it only after you register with their sites.

Delivery time: Shipping times can vary based on where an online supplier ships from. For example, a supplier in a neighboring state might get you parts in a day or so, whereas one across the country might make you wait four or five days for your order, depending on the shipping method you choose.

Labels: Suppliers typically send parts separated into many little bags, but different suppliers have different methods of labeling parts. Some don't label parts all that well, so if you aren't yet confident enough to identify one component from another, you might want to look for a supplier who is meticulous about labeling. After ordering from a few suppliers, you can determine what style of labeling works for you.

Cost: Cost is an obvious differentiator. Some companies offer price guarantees that ensure that if you find a better price, they will match it. Others impress you with all their specialty items, but they offer them at less-impressive prices. If you're ordering a lot of electronic parts, you might want to find a lower-cost resource to save a lot over time.

Part version: Be careful to check out which version of a part you're ordering. Many parts are manufactured for use on an automated assembly line with a method called *surface mounting*. These do not have pins that allow you to insert them in a breadboard.

Dimensions: Pay attention to dimensions. Some parts are so small that you can't easily handle them. Because the sizes are often listed in millimeters (which you might not be used to deciphering), you might get a component that's ⅛" x ⅛" instead of getting a component that you assumed was about ½" x ½".

Help: Need some help using the parts you buy? Some suppliers specialize in certain special interest areas and offer articles or tutorials on their Web sites. Other suppliers give you easy access to the manufacturer's datasheet, which provides useful information to help you see how to apply the part in a circuit.

Reynolds Electronics

```
http://rentron.com
```

Reynolds Electronics is a good supplier of remote control components, microcontrollers, and robot kits and parts. One feature that stands out on Reynolds

Web site is the clear and helpful project/circuit tutorials. You can also find links for educational books on topics such as robotics and microcontrollers.

Hobby Engineering

www.hobbyengineering.com

As its name suggests, Hobby Engineering is slanted toward the hobbyist. This firm carries some useful items that can prove hard to find from suppliers who cater toward the corporate customer.

Hobby Engineering offers a good supply of microcontrollers, robot kits and parts, and miscellaneous components and tools.

Jameco

www.jameco.com

We like Jameco's catalog. It's not so big that you suspect a forest was sacrificed in its making, and it has great color photos with relatively easy-to-find components.

This is one of those suppliers that offers a low price guarantee, which means that if you can show a part offered for less, the lesser price is honored.

Jameco offers lower-priced generic products as well as name brands.

Digi-Key

www.digikey.com

Digi-Key is a large distributor of components from various manufacturers, offering a huge catalog (we're talking about 1,600 pages) with lots of choices. However, the trade-off is that the catalog uses such small print that it can be hard to read. The catalog includes line drawings but no part photos. However, Digi-Key does a good job of clearly labeling parts.

Bottom line: Digi-Key might give you more choices than a smaller supplier, so you might be able to find a part you couldn't find at other suppliers although it might take some looking to find the right part.

Mouser Electronics

`www.mouser.com`

Another large distributor of components from various manufacturers is Mouser. We like how the Mouser Web site allows you to compile separate orders for different projects. You can come back as many times as you like to complete your order and make your purchase. We also like how Mouser labels parts because the labels are easy to read and provide a lot of information about the parts.

Like Digi-Key, Mouser offers a large catalog (about 1,300 pages) with lots of choices, but it can take some looking (and eye strain) to find the right part.

RadioShack

`www.radioshack.com`

Everybody knows RadioShack because there's one on every corner (almost). This electronics convenience store is pretty much like any food convenience store: convenient but expensive. If you're stuck at 4:30 on a Sunday afternoon needing a small part or two to keep working on your project, the convenience might be worth it to you.

Be aware that not all RadioShacks are equal. Some carry a much better selection of electronics parts than others.

Fry's Electronics

`www.frys.com`
`www.outpost.com`

Fry's has stores in nine states, and if you're lucky enough to live near one, you can browse its aisles in person. Fry's stocks all kinds of electronics, including a good selection of ICs.

If you live in one of the other 41 states, try Fry's online electronics store: Outpost.com. It doesn't seem to stock as many parts as in Fry's stores, but you might find some things you need there.

Electronic Goldmine

www.goldmine-elec.com

If you're in the market for bargains, try Electronic Goldmine. This online store often offers specials that can save you money. Based in Scottsdale, Arizona, Electronic Goldmine offers great help with international orders and also offers wholesale items on auction at eBay. Make a bid, and save a few bucks!

 Try clicking the Electronic Goldmine Treasure Ball for the best product deal of the day.

Furturlec

www.futurlec.com

Furturlec offers a nicely done Web site with color photos of parts that makes for easy browsing. Furturlec offers semiconductor news links and also links to recently added parts to help you keep on top of the latest and greatest as well as PCB design and manufacturing services.

With headquarters in Asia, Furturlec ships throughout the world, and its shipping times aren't that much longer than U.S. counterparts.

Maplin

www.maplin.co.uk

The British supplier Maplin has stores scattered around the U.K. and also offers an online store.

Unlike electronic gadgets and appliances that operate on different voltages in European countries, electronic components are pretty standard. Although Maplin will ship overseas, it will cost you, so unless you live in the U.K., check out U.S. suppliers first.

 To easily link to all the Web sites listed in this chapter as well as other resources, go to the author's site, BuildingGadgets, at www.buildinggadgets.com.

Chapter 16

Ten Great Electronics Resources

*I*f you have the electronics bug, you'll probably never get enough. That's why we include this chapter, which outlines some great general electronics resources.

Here we list ten — okay, a few more than ten — useful sources where you can find general information on how electronics work and also where you can get some ideas for circuits and projects. You can even chat with others who share your interests in online discussion forums.

Note that many of the projects described on these sites use much higher voltage and current levels than the projects in this book. Make sure that you have sufficient training and knowledge before tackling such projects.

Electronics Magazines

The world might live on the Internet today, but sometimes there's nothing like having a colorful magazine in hand that you can flip through while riding on the bus or eating a pastrami sandwich at the local deli. Here are some good ones to explore:

Nuts & Volts magazine

www.nutsvolts.com

This monthly magazine, which you can subscribe to for about $25 a year, uses the tagline *Everything for Electronics.* Although that might be an exaggeration, it is a good electronics hobbyist magazine with articles on a variety of types of projects, many of them microcontroller- or microprocessor-based. You're likely to find some interesting suppliers among the advertisers here as well as info on the latest technology advances and electronics products.

Everyday Practical Electronics magazine

www.epemag.wimborne.co.uk

(online version) www.epemag.com

This monthly electronics hobbyist magazine, published in the U.K., is a good choice to browse for projects using reasonably simple electronics circuits. Each issue includes lots of projects as well as reader-contributed ideas, a help desk feature, and new technology update.

You can subscribe to either a paper publication or online version. Online, you can buy back issues, subscribe to the full year for $15.99 as of this writing, or just buy a single online issue. You can also subscribe to the print magazine; prices vary depending on whether you are in the U.K. or another country.

Silicon Chip magazine

http://www.siliconchip.com.au

Here's an electronics hobbyist magazine published monthly Down Under. You can get the online version for about $59 U.S. for a year's subscription. You can get a printed version, but you'll have to fork over extra money for the international postage if you live Up Under. This Australian publication offers articles that can range from salvaging parts from your VCR to advice on wireless devices or building a vintage style radio. The Ask Silicon Chip feature advises readers on how to solve their particular problem or project challenge.

Want to get the latest reviews of hot new components from somebody you know and trust? Although we were too humble to put our own author Earl's Web site, www.buildinggadgets.com, as one of the top ten in this chapter, we hope you'll visit to get advice and tips about cool new chips and more.

Jumpstart Your Project Creativity with Circuits

A circuit is where most electronic projects start, so being able to find cool circuits is essential. Here are several online sources for circuits that offer stimulating project ideas.

Electronics Lab

www.electronics-lab.com/projects/index.html

This site is an electronic hobbyist's dream, offering schematics and explanations for several hundred electronics projects. Projects are neatly divided into categories such as Audio, Radio Frequency, Science Related, and Telephone Related.

Check out the download page for free, downloadable calculators.

Circuits for the Hobbyist

www.uoguelph.ca/~antoon/circ/circuits.htm

Run by hobbyist Tony van Roon, this site is kept up to date with over 100 schematics and very active discussion forums on topics ranging from radio projects to remote control and microcontroller-related projects.

Post your own message to get a dialog going with other people who share your interests. One nice feature of this site is an absence of advertising: It's hosted from pure enthusiasm, not greed. However, Tony can't handle non-English messages, so those are deleted.

Tony van Roon also runs a site that features tutorials on electronics plus several high-voltage projects for the adventurous (and safety minded). Go to www.uoguelph.ca/~antoon/index.htm to access this site.

Discover Circuits

www.discovercircuits.com/list.htm

This site is run by electronic engineer Dave Johnson, who states, "Electronics is my vocation, hobby and passion."

This site boasts over 11,000 electronic circuits cross-referenced into 500+ categories. Check out the featured circuit of the week for inspiration and also explore the Hot Links! to go to other sites of interest.

Links to some of Dave's other Web sites, such as Imagineering On-Line Magazine and Discover Solar Energy.com, can lead you into some other interesting materials.

Bowden's Hobby Circuits

```
http://ourworld.compuserve.com/homepages/Bill_Bowden/homepage.htm
```

This small collection of electronic circuits for the hobbyist provides over 100 circuit diagrams, links to related sites, and commercial kits and projects. In addition, you can follow links to related newsgroups and educational areas.

The JavaScript calculators collection on this site is very impressive. Here you can find calculators to help with calculating resistor color codes, RC time constants (discussed in Chapter 9), Ohm's Law, and more.

FC's Electronic Circuits

```
www.solorb.com/elect
```

This rather simply designed Web site offers interesting collection of circuits (including schematics and descriptions). What you'll find here are some rather unusual circuits that put the emphasis on practicality rather than merely electronic toys.

Most circuits are designed with discrete components that you can easily find. This site also includes an extensive list of links to other electronics sites.

Web Sites That Teach You the Ropes

Whether you need a tutorial on soldering techniques or a primer to help you understand just how electricity works, you can likely find that info on the Internet somewhere. Here are some good ties to get you started with both tutorials and discussions.

Electronics Teacher Web site

www.electronicsteacher.com

This Web site provides you with a wide range of tutorials on electronics and robotics. One nice feature is that the tutorials are ranked as beginner, intermediate, and advanced, so you can easily find the ones that meet your needs. This site also offers electronic calculators and converters as well as the ability to send an e-mail to Ask an Expert; your response will be posted in its forum.

Click the Components link to be taken to a list of suppliers and online catalogs from around the world.

The Electronics Club Web site

www.kpsec.freeuk.com

If you feel like you're a beginner in electronics, check out this site. It offers basic explanations of a range of electronics topics, from components to projects. You can also find a good table of electronics symbols, construction advice, and a good, basic soldering guide. Follow the Studying Electronics links to give yourself a good basic online electronics education.

Electronics Tutorials Web site

www.electronics-tutorials.com

Go to this site for over 120 online explanations of a variety of electronics topics. The directory of topics here makes finding what you need easy. And the Links page offers a lot of links from parts suppliers to electronics in education sites, with good explanations of what you can find there.

If English isn't your native language, you can use the translator links at the top of the home page here to view the contents in several other languages.

All About Circuits discussion forum

http://forum.allaboutcircuits.com

This is a very active discussion forum with over 5,000 members. Here you can post questions on general electronics questions, electronics textbook problems that you're stuck on, or programming for microcontrollers. If you need a break from electronics, check out the Off Topic Lounge to help you get away from it all with electronically minded colleagues.

The Electronics Lab Forum Web site (www.electronics-lab.com/forum) mentioned in the previous section also has a very active electronic discussion forum where you can post questions on circuits you are working on or questions in electronics theory.

Writing the Book on Electronics

Because we get royalties, we're big believers in books. Here are two that we think you should have in your library, in addition to *Electronics Projects For Dummies*.

- *The Art of Electronics* (Cambridge University Press, 2001) is a classic electronics reference book by Paul Horiwitz and Winfield Hill. What makes it a classic is the thorough coverage of analog and digital circuit design topics. In addition, it is well written and gives a foundation for designing your own circuits.

- *Electronics For Dummies* (Wiley, 2005) was written by Gordon McComb and Earl Boysen. (Yes, you saw that name on the cover of this book.) We happen to think that *For Dummies* books, with their easy-to-understand style, are great. We recommend this one for a good introduction to electronics theory as well as some good general information about topics such as setting up your electronics workshop and how PCB boards are made.

To easily link to all the Web sites listed in this chapter as well as other resources, go to the author's site, BuildingGadgets, at www.buildinggadgets.com.

Chapter 17

Ten Specialized Electronics Resources

In This Chapter

▶ Sounding out resources for radio-related projects

▶ Scoping out audio resources

▶ Exploring the world of robotics

*I*n Chapter 16, we provide several good resources for general electronics topics. As you work in the world of electronics, though, you might find yourself drawn to certain specialty areas, such as ham radio or robotics. Consequently, we picked a few such specialties and compiled this list of useful publications and Web sites worth looking into.

Many of the projects described on these sites use much higher voltage and current levels than the projects in this book. Make sure that you have sufficient training and knowledge before tackling such projects.

Radio

Ham radio is very much alive and well. (The original term *ham* came from early telegraphy and radio operators who were often frustrated by amateurs — called *hams* — who could jam up the system. Radio hobbyists adopted the phrase to describe themselves.) If the radio branch of electronics is your interest, check out these sites.

Ian Purdie's electronics tutorial radio design pages

http://my.integritynet.com.au/purdic

This site, designed for amateur ham radio and electronic project enthusiasts, includes tutorials and information about project kits as well as links to suppliers of ham radio equipment. You can find information on radio terminology if you're new to ham and want to sound cool. Here you can find links to books, software, and more.

QRP Quarterly

www.qrparci.org

This journal of the non-profit QRP Amateur Radio Club International is published (logically enough) four times a year by a group of hardworking volunteers. Every issue is packed with often interesting articles by some big names in the amateur radio world. *QRP Quarterly* provides a wealth of circuits, electronic projects, and schematics in a very professional format.

Australian Radio Resource Page

www.alphalink.com.au/~parkerp/project.htm

Admittedly "home-brewed," the projects offered here sometimes contain articles and supporting information, but many offer only a schematic and some notes, so you might have to figure out some things on your own. Projects range from broadcast and shortwave to marine and scanning devices. If you're a radio history buff, you can get in a time machine by visiting the archives here to read articles written way back in the 1990s.

QRP/SWL HomeBuilder

www.qrp.pops.net/default.htm

The QRP/SWL HomeBuilder site is great for short wave radio buffs, featuring many projects with clearly drawn schematics and often helpful pictures. The Junk Box feature is where the site owner posts ideas and experiments in progress, so it's an interesting way to study how a fellow ham radio devotee thinks.

There's no place like home

You also might be able to find knowledgeable people right in your own backyard:

✔ **Check out your local ham radio club.** Look for people with a range of knowledge and experience, from retired electronics engineers to "Elmers" skilled in building gadgets from what is available.

✔ **Wander into a local hobby shop.** For example, one such shop near us has several hobbyists gather around the model train tracks every Saturday. Folks like this — even though they might not have the electronics circuit knowledge you seek — will have something to share about how to build the mechanical/structural aspects of your projects.

1K3OIL

```
http://digilander.libero.it/ik3oil/menu_eng.htm
```

This site proclaims that it's devoted to amateur radio home-brewing, which translates into some interesting experimentation and idea swapping. With good color photos of projects and schematics and notes in linked PDF files, you'll get some good ideas and even contribute some of your own.

Browse through the Projects page of this site to find a really good idea of what radio enthusiasts can build when they set their minds to it.

Audio and Music

If you need music or sound effects for your projects, here are three sites that are worth a look. They range from a site devoted to guitar effects to a site on amplifier designs.

The Guitar Effects Oriented (GEO) Web Page

```
http://www.geofex.com
```

Offering a variety of guitar-related projects, the FX Projects page on this site is a good place to get started with guitar special effects.

Look at the Guitar Effects FAQ for the section Finding Guitar Effects Schematics. Check this out for a list of several other Web sites that offer information on guitar special effects.

Bob's Vacuum Tube Audio Projects Page

`www.geocities.com/TimesSquare/1965/projects.html`

Here you can find a collection of audio amplifier designs of various descriptions. If it's amp designs you need, you're likely to find a good one here. The site is a little out of date, but it's worth a visit if this is your area of interest.

Effectronics

`www.j.philpott.com/effectronics`

If you need still more do-it-yourself guitar electronic effects, head over to Effectronics. Unlike Bob's Vacuum Tube site (see the preceding section), this site is pretty much all about guitar effects. You can find design advice and schematics here as well as often lengthy and comprehensive PDF tutorials. If you've been looking for books on electronic effects for musicians, visit the Sponsors page, where you can click on a link to purchase featured books from Amazon.com.

Robotics

Whether your idea of a robot is a clanking, walking tin can from a 1950s space movie or a modern day robotic vacuum cleaner, you might be interested in building your own robots. Here are a couple of resources that can help you get started.

The BEAM Reference Library

`www.solarbotics.net/library.html`

BEAM is an approach to robotics started by a former researcher at the Los Alamos National Laboratory who now runs research and development at a toy company. BEAM robots start with simple reflexes and build on those to create mostly noncomputerized robots. Unlike processor-based robots, BEAM robots are cheap and simple, and you can build them with only the most basic skills in a matter of hours. Working with BEAM robotics can be an interesting way to quickly get down the basics of robots.

To easily link to all the Web sites listed in this chapter as well as other resources, go to the author's site, BuildingGadgets, at `www.buildinggadgets.com`.

Robot magazine

`www.botmag.com`

Even if you're not (yet) into robots, this magazine is worth looking through for electronics ideas. If you have caught the robotics bug, this magazine will have you in robot heaven. You'll find lots of color pictures of projects as well as information about robot competitions and technology advances. Want the latest robot kits? You'll find them reviewed or advertised here.

Glossary

*E*lectronics has its own jargon, just like every other discipline. Some of these terms have to do with electricity itself, such as *voltage* or *electron*. Other terms are the names of various components or tools that you work with.

Use this handy glossary if you run into an unfamiliar word or phrase.

60/40 rosin core: Solder used in working with electronics that contains 60 percent tin and 40 percent lead (within a few percentage points) with a core of rosin flux.

alternating current (AC): Current in which there is a change in the direction in which electrons flow. *See also* direct current (DC).

amplitude: How much voltage is in an electrical signal.

anode: Positive terminal of a diode. *See also* cathode.

auto-ranging: A feature that some multimeters offer that automatically sets a test range. *See also* multimeter.

AWG (American Wire Gauge): *See* wire gauge.

bandwidth: With an oscilloscope, the highest frequency signal that you can test with any reliability, measured in megahertz (MHz).

battery: A power source that uses electrochemical reaction to create a positive voltage at one terminal and a negative voltage at another. Two different types of metal are placed in a type of chemical to produce the power.

biasing: Applying a small amount of voltage to the base of a transistor. This partially turns on the transistor.

bipolar: A common IC type. *See also* integrated circuit.

breadboard: Also referred to as a *prototyping board* or *solderless breadboard*. Plastic boards in a variety of shapes, styles, and sizes that have columns of holes. A line of metal connects these holes electrically. By plugging components into these holes and connecting them with wire, you can build a circuit. *See also* soldered breadboard.

bus: A connection point.

cable: Sets of two or more wires that are protected by an outer insulation layer. The common power cord is an example of a cable.

capacitance: The ability to store electrons. You measure capacitance in farads.

capacitor: A component in a circuit providing the property of capacitance. *See also* electrolytic capacitor, ceramic capacitor, tantalum capacitor.

cathode: The negative terminal of a diode. *See also* anode.

ceramic capacitor: One of the most common types of capacitor. This type is used for smaller values of capacitance. *See also* capacitor.

circuit: Wires that connect components in such a way that a current flows through the components and returns to the source.

closed circuit: A circuit with connected wires, so that current is able to flow. *See also* open circuit.

closed position: The position of a switch in which current can flow. *See also* open position.

CMOS: A type of IC that is very sensitive to ESD. *See also* integrated circuit, static electricity.

cold solder joints: A bad joint. These happen when solder doesn't flow properly around metal parts.

commutator: A device that changes the direction of electric current in a motor or generator.

components: The parts you use in electronics projects, such as a capacitor or resistor.

conductor: A material through which electricity moves freely.

connector: Metal or plastic receptacles on equipment into which you can fit cable ends.

continuity: A multimeter test of whether a circuit is intact between two points. *See also* multimeter.

current: The flow of an electrical charge.

cycle: That portion of an AC waveform where the voltage goes from its lowest point to the highest point and back again. This cycle repeats again and again until you turn off the signal.

decoder: An IC that takes information transmitted by either infrared or radio remote signal and translates it into output signals. *See also* encoder, integrated circuit.

desolder pump: A piece of equipment that sucks up excess solder by using a vacuum.

diode: A device that limits the flow of current to one direction, thereby converting alternating current to direct current.

direct current (DC): A type of current in which the electrons move only from the negative terminal through the wires to the positive terminal in one direction. A battery generates direct current.

double-pole, double-throw switch (DPDT): A type of switch with two input connections and four output connections.

double-pole, single-throw switch (DPST): A type of switch with two input connections and two output connections.

double-pole switches: A type of switch with two input connections.

DPDT: *See* double-pole, double-throw switch.

DPST: *See* double-pole, single-throw switch.

dual inline package (DIP): A commonly used integrated circuit plastic package; used in breadboards and circuit boards used by hobbyists.

electricity: The movement of electrons through a conductor.

electrolytic capacitor: One of the most common types of capacitor. This type is used for larger values of capacitance. *See also* capacitor.

electromagnet: A type of wire coiled around a piece of metal (usually an iron bar). When current runs through the wire, the metal is magnetized. Shut off the current, and the metal loses its magnetic quality.

electromotive force: An attractive force which you measure in volts. This force exists between positive and negative charges.

electron: A negatively charged particle. *See also* proton.

encoder: An IC that codes information to be transmitted by either infrared or radio remote signal. *See also* decoder, integrated curcuit.

ESD (electrostatic discharge): *See* static electricity.

farad: The unit of measurement for capacitance. (A microfarad is one millionth of a farad.) *See also* capacitor.

flathead: A type of screw with a flat head and single slot; the screwdriver used with a flathead screw.

flux: A waxy substance used to make molten solder flow around components and wire, assuring a good joint.

frequency: A way to measure how often an AC signal repeats. The symbol for frequency is *f.*

gain: The amount that a signal is amplified. To calculate this, divide the voltage of the signal coming out by the voltage of the incoming signal.

gauge: *See* wire gauge.

ground: A connection in a circuit used as a reference for 0 (zero) volts.

H-bridge: A circuit or IC used to control the power to DC motors. *See also* intergrated circuit.

heat sink: A piece of metal used to protect components. The sink is attached to the component and draws off heat to prevent destroying the component.

hertz (Hz): The measurement used for the number of cycles per second that occur in alternating current.

high pass filter: A circuit that allows signals above a certain frequency to pass through. *See also* low pass filter.

high signal: In digital electronics, a signal at a value of higher than 0 (zero) volts.

I: The symbol used for current.

IC: *See* integrated circuit.

impedence: In an electrical current, a measurement of opposition to the flow of alternating current.

inductance: Ability to store energy in a magnetic field (measured in Henries).

inductors: Components that provide the ability to store energy in a magnetic field for a circuit.

infrared temperature sensor: A type of sensor that measures temperature electrically.

insulator: A substance that electrons are unable to move through freely.

integrated circuit (IC): A component (in the form of a chip) that contains several smaller components, such as resistors, capacitors, and transistors.

inverter: A type of logic gate with a single input. *See also* logic gate.

inverting mode: What happens when an op amp flips an input signal to produce an output signal. *See also* operational amplifier.

jack: A type of connector. *See also* connector.

kohm: 1000 ohms. *See also* ohm.

live circuit: A circuit with voltage applied.

logic gate: An integrated circuit that uses input values to determine output value based on certain rules.

low pass filter: A circuit that allows signals below a certain frequency to pass through. *See also* high pass filter.

low signal: In digital electronics, a signal at or near 0 (zero) volts.

lug: A metal protrusion, usually with a hole in the center, into which you can feed wire and solder it to various components.

microcontroller: A programmable circuit.

multimeter: A testing device that measures things such as voltage, resistance, and amperage.

n-type semiconductor: A semiconductor to which contaminants are added. This causes it to have more electrons than a pure semiconductor.

ohm: A unit of resistance. The symbol for ohm is Ω. *See also* resistance.

Ohm's Law: The equation that you use to calculate voltage, current, resistance, or power.

open circuit: A circuit in which a wire is disconnected. Therefore, no current is able to flow. *See also* closed circuit.

open position: A switch position that stops current from flowing. *See also* closed position.

operational amplifier: Also called *op amp.* An integrated circuit that contains transistors and other components. An op amp provides uniform amplification over a wider range of frequencies than a single-transistor amplifier.

oscillator: A circuit that generates waveforms. *See also* waveform.

oscilloscope: An electronic device used to measure voltage, frequency, and other parameters for waveforms.

p-type semiconductor: A semiconductor to which contaminants are added that cause it to contain fewer electrons than a pure semiconductor.

pad: Contact points on a circuit board used to connect components.

Phillips: Both a screw with a plus-shaped (+) slot in its head and the screwdriver that you use with that type of screw.

pn junction: The interface of two regions that contain boron and phosphorus adjacent in a semiconductor. Transistors and diodes contain pn junctions. *See also* transistor, diode.

potentiometer: A variable resistor that allows for continual adjustment of resistance. This adjustment can range from virtually 0 (zero) ohms to a maximum value.

power: The measure in watts of the amount of work that electric current does while running through an electrical component.

proton: A positively charged particle. *See also* electron.

prototyping board: *See* breadboard.

pulse: A signal that rapidly alternates between high and low.

pulse width modulation: A way to control the speed of a motor by turning voltage on and off in quick pulses. When the on intervals are longer, the motor goes faster.

R: The symbol for resistance.

RC (resistance/capacitance) time constant: A formula to calculate the time required to charge a capacitor to two-thirds or discharge it to one-third of its capacity.

relay: A device that performs like a switch. It closes or opens a circuit, depending on the voltage that you supply.

resistance: The measurement of the ability of electrons to move through any material.

resistor: A component of a circuit that reduces the amount of electrons that flow through the circuit.

rosin flux remover: An after-soldering cleaner that removes any remaining flux so that it doesn't oxidize your circuit.

schematic: A drawing that shows how components in a circuit are connected.

semiconductor: A material, such as silicon, that has some of the properties of both conductors and insulators.

semiconductor temperature sensors: A type of sensor that measures temperature electrically.

sensors: Electronic components that sense a particular condition, such as heat or light.

series circuit: A circuit in which current runs through each component in sequence.

short circuit: What happens when two wires are connected and current goes through them, resulting in the circuit not being completed.

sine wave: An signal that switches from a high voltage to a low voltage, back to the high voltage, and then repeats this cycle with a smooth *sinesoidal* (curvy) waveform until it's shut off.

single-pole, double-throw switch (SPDT): A type of switch in which one wire goes into the switch and two wires leave the switch.

single-pole switches: A type of switch with one input wire.

slide switch: A switch that you slide forward or backward to turn an electronic device on or off.

solar cell: A semiconductor that generates a current when exposed to light.

solder sucker: A tool used to remove excess solder. The sucker uses a spring-loaded vacuum.

solder wick: Also called *solder braid.* A device that you use to remove difficult-to-reach solder. The solder wick is a copper braid that absorbs solder more easily than the tin plating on most components and printed circuit boards.

soldered breadboard: A breadboard on which you solder components. *See also* breadboard.

soldering: The method that uses small globs of molten metal — *solder* — to hold components together.

soldering iron: *See* soldering pencil.

soldering pencil: A wand-like tool used to apply solder.

solderless breadboard: *See* breadboard.

solid wire: A single-strand wire. *See also* stranded wire.

SPDT: *See* single-pole, double-throw switch.

spike: *See* voltage spike.

square wave: An signal that switches from a high voltage to a low voltage, back to the high voltage, and then repeats this cycle with a square-shaped waveform until it's shut off.

static electricity: A form of current that remains within an insulating body after you remove the power source. Lightning is an example of static electricity.

stranded wire: Two or three small bundles of very fine wires wrapped within insulation. *See also* solid wire.

stray capacitance: What happens when electric fields occur between wires or leads in a circuit because they are placed too close together. In this condition, energy is stored unintentionally.

tantalum capacitor: A type of capacitor used when the circuit can't accommodate a large range of capacitance value variation. *See also* capacitor.

terminal: A piece of metal to which you hook up wires (for example, a battery terminal).

thermistor: A resistor that changes its resistance value when the temperature changes.

thermocouple: A sensor that measures temperature electrically.

tinning: The process of heating a soldering tool to full temperature and applying a small amount of solder to the tip. This prevents solder from sticking to the tip.

tolerance: The variation in the value of a component due to the manufacturing process that is allowable, typically expressed as a range.

traces: On a circuit board, the wires that run between pads to electrically connect components.

transistor: A semiconductor that controls the flow of electric current.

V: The symbol for voltage. Also can represented by E.

variable capacitor: A capacitor that includes two or more metal plates that are separated by air. If you turn the knob, you change the capacitance of a device. *See also* capacitor.

variable coil: A coil of wire that surrounds a movable metal slug. When you turn the slug, you change or vary the inductance of the coil.

variable resistor: *See* potentiometer.

voltage: Attractive force between positive and negative charges.

voltage divider: The voltage drops in a circuit that produce voltage lower than the supply voltage at certain points in the circuit.

voltage drop: The lowering of voltage that occurs when voltage pulls electrons through resistors (or any other component) and the component uses up a portion of the voltage.

voltage spike: A brief increase in voltage.

watt hour: A unit of measure for energy; the ability of a device or circuit to do work.

waveform: Voltage fluctuations such as those that appear in a sine wave or square wave. *See also* oscilloscope, sine wave, square wave.

wire: A long strand of metal, usually made of copper, used to make connections in electronics projects. Electrons travel through the wire, conducting electricity.

wire clips: An adhesive-backed piece of plastic with a clip used to secure wires.

wire gauge: A system used to measure the diameter of wire.

Index

• *C* •

• T •

• *X* •

• *Z* •

Notes

Notes

BUSINESS, CAREERS & PERSONAL FINANCE

Grant Writing FOR DUMMIES

0-7645-5307-0

Home Buying FOR DUMMIES

0-7645-5331-3 *†

Also available:
- Accounting For Dummies †
  0-7645-5314-3
- Business Plans Kit For Dummies †
  0-7645-5365-8
- Cover Letters For Dummies
  0-7645-5224-4
- Frugal Living For Dummies
  0-7645-5403-4
- Leadership For Dummies
  0-7645-5176-0
- Managing For Dummies
  0-7645-1771-6
- Marketing For Dummies
  0-7645-5600-2
- Personal Finance For Dummies *
  0-7645-2590-5
- Project Management For Dummies
  0-7645-5283-X
- Resumes For Dummies †
  0-7645-5471-9
- Selling For Dummies
  0-7645-5363-1
- Small Business Kit For Dummies *†
  0-7645-5093-4

HOME & BUSINESS COMPUTER BASICS

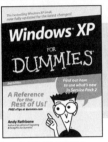

Windows XP FOR DUMMIES

0-7645-4074-2

Excel 2003 ALL-IN-ONE DESK REFERENCE FOR DUMMIES

0-7645-3758-X

Also available:
- ACT! 6 For Dummies
  0-7645-2645-6
- iLife '04 All-in-One Desk Reference
  For Dummies
  0-7645-7347-0
- iPAQ For Dummies
  0-7645-6769-1
- Mac OS X Panther Timesaving
  Techniques For Dummies
  0-7645-5812-9
- Macs For Dummies
  0-7645-5656-8
- Microsoft Money 2004 For Dummies
  0-7645-4195-1
- Office 2003 All-in-One Desk Reference
  For Dummies
  0-7645-3883-7
- Outlook 2003 For Dummies
  0-7645-3759-8
- PCs For Dummies
  0-7645-4074-2
- TiVo For Dummies
  0-7645-6923-6
- Upgrading and Fixing PCs For Dummies
  0-7645-1665-5
- Windows XP Timesaving Techniques
  For Dummies
  0-7645-3748-2

FOOD, HOME, GARDEN, HOBBIES, MUSIC & PETS

Feng Shui FOR DUMMIES

0-7645-5295-3

Poker FOR DUMMIES

0-7645-5232-5

Also available:
- Bass Guitar For Dummies
  0-7645-2487-9
- Diabetes Cookbook For Dummies
  0-7645-5230-9
- Gardening For Dummies *
  0-7645-5130-2
- Guitar For Dummies
  0-7645-5106-X
- Holiday Decorating For Dummies
  0-7645-2570-0
- Home Improvement All-in-One
  For Dummies
  0-7645-5680-0
- Knitting For Dummies
  0-7645-5395-X
- Piano For Dummies
  0-7645-5105-1
- Puppies For Dummies
  0-7645-5255-4
- Scrapbooking For Dummies
  0-7645-7208-3
- Senior Dogs For Dummies
  0-7645-5818-8
- Singing For Dummies
  0-7645-2475-5
- 30-Minute Meals For Dummies
  0-7645-2589-1

INTERNET & DIGITAL MEDIA

Digital Photography FOR DUMMIES

0-7645-1664-7

Starting an eBay Business FOR DUMMIES

0-7645-6924-4

Also available:
- 2005 Online Shopping Directory
  For Dummies
  0-7645-7495-7
- CD & DVD Recording For Dummies
  0-7645-5956-7
- eBay For Dummies
  0-7645-5654-1
- Fighting Spam For Dummies
  0-7645-5965-6
- Genealogy Online For Dummies
  0-7645-5964-8
- Google For Dummies
  0-7645-4420-9
- Home Recording For Musicians
  For Dummies
  0-7645-1634-5
- The Internet For Dummies
  0-7645-4173-0
- iPod & iTunes For Dummies
  0-7645-7772-7
- Preventing Identity Theft For Dummies
  0-7645-7336-5
- Pro Tools All-in-One Desk Reference
  For Dummies
  0-7645-5714-9
- Roxio Easy Media Creator For Dummies
  0-7645-7131-1

SPORTS, FITNESS, PARENTING, RELIGION & SPIRITUALITY

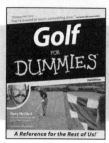

0-7645-5146-9

0-7645-5418-2

Also available:
- Adoption For Dummies
  0-7645-5488-3
- Basketball For Dummies
  0-7645-5248-1
- The Bible For Dummies
  0-7645-5296-1
- Buddhism For Dummies
  0-7645-5359-3
- Catholicism For Dummies
  0-7645-5391-7
- Hockey For Dummies
  0-7645-5228-7

- Judaism For Dummies
  0-7645-5299-6
- Martial Arts For Dummies
  0-7645-5358-5
- Pilates For Dummies
  0-7645-5397-6
- Religion For Dummies
  0-7645-5264-3
- Teaching Kids to Read For Dummies
  0-7645-4043-2
- Weight Training For Dummies
  0-7645-5168-X
- Yoga For Dummies
  0-7645-5117-5

TRAVEL

0-7645-5438-7

0-7645-5453-0

Also available:
- Alaska For Dummies
  0-7645-1761-9
- Arizona For Dummies
  0-7645-6938-4
- Cancún and the Yucatán For Dummies
  0-7645-2437-2
- Cruise Vacations For Dummies
  0-7645-6941-4
- Europe For Dummies
  0-7645-5456-5
- Ireland For Dummies
  0-7645-5455-7

- Las Vegas For Dummies
  0-7645-5448-4
- London For Dummies
  0-7645-4277-X
- New York City For Dummies
  0-7645-6945-7
- Paris For Dummies
  0-7645-5494-8
- RV Vacations For Dummies
  0-7645-5443-3
- Walt Disney World & Orlando For Dummies
  0-7645-6943-0

GRAPHICS, DESIGN & WEB DEVELOPMENT

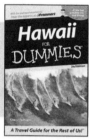

0-7645-4345-8

0-7645-5589-8

Also available:
- Adobe Acrobat 6 PDF For Dummies
  0-7645-3760-1
- Building a Web Site For Dummies
  0-7645-7144-3
- Dreamweaver MX 2004 For Dummies
  0-7645-4342-3
- FrontPage 2003 For Dummies
  0-7645-3882-9
- HTML 4 For Dummies
  0-7645-1995-6
- Illustrator CS For Dummies
  0-7645-4084-X

- Macromedia Flash MX 2004 For Dummies
  0-7645-4358-X
- Photoshop 7 All-in-One Desk Reference For Dummies
  0-7645-1667-1
- Photoshop CS Timesaving Techniques For Dummies
  0-7645-6782-9
- PHP 5 For Dummies
  0-7645-4166-8
- PowerPoint 2003 For Dummies
  0-7645-3908-6
- QuarkXPress 6 For Dummies
  0-7645-2593-X

NETWORKING, SECURITY, PROGRAMMING & DATABASES

0-7645-6852-3

0-7645-5784-X

Also available:
- A+ Certification For Dummies
  0-7645-4187-0
- Access 2003 All-in-One Desk Reference For Dummies
  0-7645-3988-4
- Beginning Programming For Dummies
  0-7645-4997-9
- C For Dummies
  0-7645-7068-4
- Firewalls For Dummies
  0-7645-4048-3
- Home Networking For Dummies
  0-7645-42796

- Network Security For Dummies
  0-7645-1679-5
- Networking For Dummies
  0-7645-1677-9
- TCP/IP For Dummies
  0-7645-1760-0
- VBA For Dummies
  0-7645-3989-2
- Wireless All In-One Desk Reference For Dummies
  0-7645-7496-5
- Wireless Home Networking For Dummies
  0-7645-3910-8

HEALTH & SELF-HELP

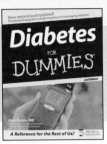

0-7645-6820-5 *†

0-7645-2566-2

Also available:

- Alzheimer's For Dummies
  0-7645-3899-3
- Asthma For Dummies
  0-7645-4233-8
- Controlling Cholesterol For Dummies
  0-7645-5440-9
- Depression For Dummies
  0-7645-3900-0
- Dieting For Dummies
  0-7645-4149-8
- Fertility For Dummies
  0-7645-2549-2

- Fibromyalgia For Dummies
  0-7645-5441-7
- Improving Your Memory For Dummies
  0-7645-5435-2
- Pregnancy For Dummies †
  0-7645-4483-7
- Quitting Smoking For Dummies
  0-7645-2629-4
- Relationships For Dummies
  0-7645-5384-4
- Thyroid For Dummies
  0-7645-5385-2

EDUCATION, HISTORY, REFERENCE & TEST PREPARATION

0-7645-5194-9

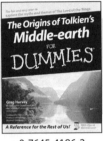

0-7645-4186-2

Also available:

- Algebra For Dummies
  0-7645-5325-9
- British History For Dummies
  0-7645-7021-8
- Calculus For Dummies
  0-7645-2498-4
- English Grammar For Dummies
  0-7645-5322-4
- Forensics For Dummies
  0-7645-5580-4
- The GMAT For Dummies
  0-7645-5251-1
- Inglés Para Dummies
  0-7645-5427-1

- Italian For Dummies
  0-7645-5196-5
- Latin For Dummies
  0-7645-5431-X
- Lewis & Clark For Dummies
  0-7645-2545-X
- Research Papers For Dummies
  0-7645-5426-3
- The SAT I For Dummies
  0-7645-7193-1
- Science Fair Projects For Dummies
  0-7645-5460-3
- U.S. History For Dummies
  0-7645-5249-X

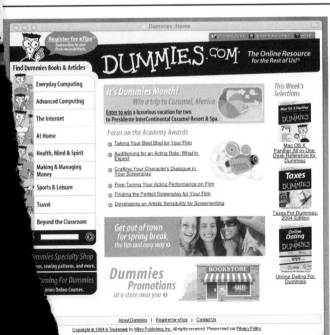

Get smart @ dummies.com®

- **Find a full list of Dummies titles**
- **Look into loads of FREE on-site articles**
- **Sign up for FREE eTips e-mailed to you weekly**
- **See what other products carry the Dummies name**
- **Shop directly from the Dummies bookstore**
- **Enter to win new prizes every month!**

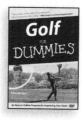